THE URBAN MOMENT

URBAN AFFAIRS ANNUAL REVIEWS

A series of reference volumes discussing programs, policies, and current developments in all areas of concern to urban specialists.

SERIES EDITORS

The **Urban Affairs Annual Reviews** presents original theoretical, normative, and empirical work on urban issues and problems in regularly published volumes. The objective is to encourage critical thinking and effective practice by bringing together interdisciplinary perspectives on a common urban theme. Research that links theoretical, empirical, and policy approaches and that draws on comparative analyses offers the most promise for bridging disciplinary boundaries and contributing to these broad objectives. With the help of an international advisory board, the editors will invite and review proposals for **Urban Affairs Annual Reviews** volumes that incorporate these objectives. The aim is to ensure that the **Urban Affairs Annual Reviews** remains in the forefront of urban research and practice by providing thoughtful, timely analyses of cross-cutting issues for an audience of scholars, students, and practitioners working on urban concerns throughout the world.

INTERNATIONAL EDITORIAL ADVISORY BOARD

RECENT VOLUMES

34 **ECONOMIC RESTRUCTURING AND POLITICAL RESPONSE**
Edited by Robert A. Beauregard

35 **CITIES IN A GLOBAL SOCIETY**
Edited by Richard V. Knight and Gary Gappert

36 **GOVERNMENT AND HOUSING: Developments in Seven Countries**
Edited by Willem van Vliet-- and Jan van Weesep

38 **BIG CITY POLITICS IN TRANSITION**
Edited by H. V. Savitch and John Clayton Thomas

40 **THE PENTAGON AND THE CITIES**
Edited by Andrew Kirby

41 **MOBILIZING THE COMMUNITY: Local Politics in the Era of the Global City**
Edited by Robert Fisher and Joseph Kling

42 **GENDER IN URBAN RESEARCH**
Edited by Judith A. Garber and Robyne S. Turner

43 **BUILDING THE PUBLIC CITY: The Politics, Governance, and Finance of Public Infrastructure**
Edited by David C. Perry

44 **NORTH AMERICAN CITIES AND THE GLOBAL ECONOMY: Challenges and Opportunities**
Edited by Peter Karl Kresl and Gary Gappert

45 **REGIONAL POLITICS: America in a Post-City Age**
Edited by H. V. Savitch and Ronald K. Vogel

46 **AFFORDABLE HOUSING AND URBAN REDEVELOPMENT IN THE UNITED STATES**
Edited by Willem van Vliet--

47 **DILEMMAS OF URBAN ECONOMIC DEVELOPMENT: Issues in Theory and Practice**
Edited by Richard D. Bingham and Robert Mier

48 **STATE DEVOLUTION IN AMERICA: Implications for a Diverse Society**
Edited by Lynn A. Staeheli, Janet E. Kodras, and Colin Flint

49 **THE URBAN MOMENT: Cosmopolitan Essays on the Late-20th-Century City**
Edited by Robert A. Beauregard and Sophie Body-Gendrot

THE URBAN MOMENT

COSMOPOLITAN ESSAYS ON THE LATE-20TH-CENTURY CITY

ROBERT A. BEAUREGARD
SOPHIE BODY-GENDROT
EDITORS

URBAN AFFAIRS ANNUAL REVIEWS

Sage Publications, Inc.
International Educational and Professional Publisher
Thousand Oaks ■ London ■ New Delhi

For information:

Sage Publications, Inc.
2455 Teller Road
Thousand Oaks, California 91320
E-mail: order@sagepub.com

Sage Publications Ltd.
6 Bonhill Street
London EC2A 4PU
United Kingdom

Sage Publications India Pvt. Ltd.
M-32 Market
Greater Kailash I
New Delhi 110 048 India

Printed in the United States of America

Library of Congress Cataloging-in-Publication Data

Main entry under title:
The urban moment: Cosmopolitan essays on the late-20th-century city / [edited] by Robert A. Beauregard and Sophie Body-Gendrot.
p. cm.—(Urban affairs annual reviews; v. 49)
Includes bibliographical references.
ISBN 0-7619-1484-6 (cloth: alk. paper)
ISBN 0-7619-1485-4 (pbk.: alk. paper)
1. Cities and towns. 2. City and town life. 3. Civilization, Modern—1950. 4. Postmodernism. I. Beauregard, Robert A. II. Body-Gendrot, Sophie. III. Series.
HT151 .U674 1999
307.76—dc21 99-6476

This book is printed on acid-free paper.

99 00 01 02 03 04 05 7 6 5 4 3 2 1

Acquiring Editor: Catherine Rossbach
Editorial Assistant: Caroline Sherman
Production Editor: Sanford Robinson
Editorial Assistant: Nevair Kabakian
Typesetter: Lynn Miyata
Cover Designer: Ravi Balasuriya

Contents

Part III: Civic Engagement

Part IV: Prescriptive Visions

From the Series Editors

It is with equal measures of excitement and sadness that we introduce this volume of the Urban Affairs Annual Reviews. We are excited because, from the earliest stages of planning this volume, we wanted this to be a special book, and it is. *The Urban Moment* brings together some of the most influential and distinguished urban scholars of the late 20th century to focus our attention on what editors Robert Beauregard and Sophie Body-Gendrot describe as the "unstable" nature of the city "as an object of thought and action." The question we posed to the editors of the volume was this: Given the disarray of urban thought in particular and prescriptive urban action in general, whither the city? As this century, indeed millennium, comes to a close, the world is forever changed where more than half the world's population lives in urban settlements. The increasing regularity of urban existence, however, has not made it any less problematic. If anything, the authors of this volume suggest, the opposite is true: The state of cities is even more uncertain due to "a world becoming simultaneously disarticulated and rearticulated under the onslaught of corporate globalism, ethnic social movements, state violence, massive waves of immigration and intellectual upheaval." This is not a volume of urban theory that hews to the mainstream. Its contributors engage directly this "urban moment," offering essays that address the importance of urban settlement and the fragility of the democratic city in the face of the "fluidity, discontinuity, and uncertainty" brought on by the multitude of global, economic, political, and cultural forces "we have not yet learned to master."

By the end of the volume, contributors from three continents will have traveled over risky ground, offering a full array of historical and sociospatial perspectives. These essays are at once clearly representative of urbanism in its various spatial and cultural manifestations but not so particular as to avoid the globalization of urbanism that has caused some geographers to muse over the fact that the "urban moment" is one of "global cities" in an "urban world." Such "urban ubiquity" does not overcome some contributors' desire to theorize prescriptive visions of the good city and to offer processual conceptualizations of a (re)new(ed) civil society.

As a transdisciplinary engagement of an activist urban theory—one that "dares those who are . . . dissatisfied" with the scholarship of the contemporary city to actively do more—this volume does exactly what a good volume of the UAARs should do. And this brings us to what makes us so sad about this volume: It is the last one. After over 30 years and nearly 50 volumes, the Urban Affairs Annual Reviews series, often simply known as the "Annuals," comes to an end here. When we first began to plan for this volume of the UAARs, we conceptualized it as an anniversary volume that would celebrate the publication of three decades of wide-ranging books of original social science and policy analysis on contemporary urbanism. Instead, *The Urban Moment* has become a capstone to what has been the longest and most well-known international book series of its kind dedicated to the study of the city.

The UAARs, along with the journal *Urban Affairs Quarterly,* was conceived by Sara Miller McCune, the co-founder and president of Sage Publications in 1965, during days of urban unrest, protest over the Vietnam war, and a growing consensus that the condition of cities, in the United States and elsewhere, demanded concerted attention. Miller McCune and her husband George McCune patterned Sage Publications after the Free Press by creating a publishing house organized around topics of social analysis with immediacy and relevance. They sought to produce analytic volumes, handbooks, and journals reflective of the social, political, economic, and professional changes of modern life.

"At that time academic presses were very disciplinary," says Miller McCune, "publishing books by a scholar in a discipline for the members of that discipline. The problem was that publications in the social sciences did not actively reflect the urban world—they didn't cross disciplines to study what was fully happening in, say, cities." Indeed, there did not appear to be an interdisciplinary market for the books either. The UAARs and *Urban Affairs Quarterly* were meant to correct this by creating an ongoing journal and book series of interdisciplinary social science of the city, thereby staking out a place for those interested in cities, in social policy, and in social science to study and publish.

With the advice of academics like Marilyn Gittell, Morris Janowitz, Leo Schnore, and Scott and Ann Greer, the concept of an annual series on urban scholarship and policymaking was framed by the McCunes and their early academic editors, and the first volume, edited by sociologist Leo Schnore and urban planning scholar Henry Fagin, appeared in 1967. The organization of *Urban Research and Policy Planning* set the agenda of the series from the beginning. The first half of the volume concentrated on the disciplines of social science and the way urbanism is captured in the academy. The second half of the book addressed "policy planning" or the conditions of urban need that cut "across disciplinary lines" and demanded "action-oriented research." This transdisciplinary blend of academic research informed by and directed toward the conditions of urban life both in the United States and around the world became the increasingly complex and rich contribution of scholarly activity one came to expect from each volume of the Annuals.

Over the years, the Annuals became in Miller McCune's words a "type of virtual community of interdisciplinary scholarly study directed at social critique and action." The Annuals started out as one large volume, published once a year and intended to quickly produce and disseminate social science research on a topic that might have an impact on policy. As the topics multiplied, the Annuals soon became semiannuals. But the goal remained the same: to produce books on urban issues in a timely fashion so that important ideas and analyses could be quickly disseminated to libraries, universities and policy makers. This network of universities and libraries has shrunk in recent years—the "virtual community" having fractured into a broad array of urban policy journals, university-press-based book series, and a growing number of public and private policy, association and journalistic Web sites and listservs.

In short, the UAARs are no longer that singularly special site of virtual communication. But, along the way, almost anyone interested in the study of the city could claim membership as editors, contributors, editorial board members, and most importantly as readers in a "community of Annuals" that has published some of the most interesting and transformative urban moments of the last half of the 20th century. The 49 books in the series have been edited by 67 volume editors from 72 distinct urban specialties and subdisciplines. The International Board of Editorial Advisors has included 58 distinguished scholars from 14 different countries. Perhaps most impressive is the fact that 756 urbanists, representing 263 self-described areas of specialization, contributed 690 chapters to the 49 volumes during the three decades of the series' existence. In short, there never has been an urban book series with quite the disciplinary reach or the international and scholarly scope of the UAARs.

Frankly, although the Annuals were highly representative of all areas of urban study, the quality and contributions of the series also mirrored quite well the ebb and flow of relevance of urban scholarship to the academy and practitioners more generally. During the series' 30 years, the contributions to the Annuals reflected treatments of the urban that were mostly vibrant but occasionally lackluster. Most often, a volume of the UAARs served as a source of critique and renewal for the study of the city. In fact, the last paragraph of Body-Gendrot and Beauregard's introduction to this volume goes a long way toward summing up what the series, at its best, was all about: challenging scholars and policy-makers to do better. The essays in *The Urban Moment* mount a clear critique of, as well as a challenge to, the confusion presently found in urban scholarship. Importantly, they also dare those who are dissatisfied to do better, be more insightful, think differently. They write, "We are witnessing a turning point in urban theory. Everywhere the lack of new ideas in the analysis of the city has been part of the urban crisis. The future of the democratic city is fragile indeed. . . . Although we do not know what directions eventually will prove fruitful or what destinations will be achieved, we nonetheless offer this book as both a documentation of that search and a dare to those who still are dissatisfied." We concur.

It is thus a bittersweet exercise to introduce this very important final volume at the same time that we say good-bye to the venue that enabled so much debate, ferment, and new ideas about how to think about and address the challenges of the changing city. The UAARs, in a dense cluster on the bookshelves of libraries and scholars throughout the world, will continue to exist as testimony to one of the most significant areas of scholarly and policy concern at the end of the 20th century: the city.

Sallie Marston, *Tucson, Arizona*
David C. Perry, *Chicago, Illinois*

Acknowledgments

We acknowledge Sallie Marston and David Perry for the role they played in the development of this book. It was their idea to mark a milestone in the evolution of the Urban Affairs Annual Reviews with a book on "the city." It also was their idea to ask us to be co-editors. They offered an intriguing proposition, and we accepted. Time passed, and they also contributed helpful editorial suggestions.

Thanks also are due to Catherine Rossbach, acquisitions editor at Sage Publications. Catherine's patience and friendly support were greatly appreciated.

Without the Internet facilities of the New School for Social Research and the University of Paris VII–Jussieu, our task would have been much more difficult. Overall, the hidden subvention of our respective institutions should not be overlooked.

Finally, we thank our contributors—busy and accomplished scholars with many demands on their time and many requests for their presence. They made a commitment to us and delivered. The result is a unique collection of readings by the major theorists in urban studies. What praise this book receives belongs to them.

Robert A. Beauregard
New York City

Sophie Body-Gendrot
Paris

Introduction

1

Imagined Cities, Engaged Citizens

SOPHIE BODY-GENDROT
and ROBERT A. BEAUREGARD

In Italo Calvino's wonderful book, *Invisible Cities,* Marco Polo entertains the Kublai Khan with stories of cities of imagination and engagement. Marco Polo's cities are elusive yet everywhere; all cities are one, and any one city contains them all. His stories explore the vertigo of urban differences and the search for the ever-advancing urban moment (Calvino, 1974).

Calvino's poetic musings problematize the city. They destabilize its meanings, doing so not to reject the city but rather to find ways in which to embrace its dizzying possibilities.

For urban scholars in the late 20th century, Marco Polo's stories probe to the core of their work. In a world becoming simultaneously disarticulated and rearticulated under the onslaught of corporate globalization, ethnic social movements, state violence, massive waves of immigration, and intellectual upheaval, the city itself has become unstable as an object of thought and action.

One consequence, and correlate, of this moment is the distrust of urban theory as meta-narrative—"totalizing," its detractors proclaim. Urban scholars avoid staking a defining perspective that precludes others or resists modification. They wander across disciplines in search of compelling explanations and even entertain multiple interpretations. Ambiguity and complexity take the place of grand theory's traditional infatuation with the formal, the unequivocal, and the parsimonious. The Chicago School is shunned, and even the "Los Angeles School" (see Steven Flusty and Michael Dear in Chapter 2 of this volume) dances gingerly across its own stage, attesting not to a final statement but rather to a representation in flux. Skepticism reigns (Toulmin, 1990).

The city is assumed to mean different things to different people. At the extreme, no one point of view is considered less or more equal, worse or better, than another. Multiple points of view are celebrated. The city is presumed to be everywhere, not as a physical form but rather as a social fact that pervades all societies. The elusive fixity of modernism seems to have struck with a vengeance (Berman, 1982). Nevertheless, many of these theorists search for a more civil, more tolerant, more democratic city and thereby imply that coexisting, diverse representations do not preclude common understandings that enable people to live together, with strangers, in cities (Bohman, 1996; Young, 1990).

As the disarticulating forces of late-20th-century political and economic arrangements gain ascendancy and the forces of integration are channeled more and more to exclusionary ends, creating this common ground seems less attainable. Under these conditions, it is easier and more comfortable to shift into the realm of idealized representation rather than to engage a messy reality collectively. That talking about the democratic city is not the same as creating it need not be belabored. Justice, equality, empathy, and tolerance are more than discursive acts, even as their meanings evaporate in the absence of representation. For intellectuals, of course, talk is action. By shaping meaning and disclosing new frames of understanding, their intent is to change minds and hearts. These changes enable people to connect with the city, or not, and are the basis for democracy and justice (Merrifield & Swyngedouw, 1997).

The essays in this volume reflect on these themes. They explore alternative ways of representing the contemporary city, the city's social and physical disarticulation and rearticulation (particularly as it relates to globalization), the prospects for democracy and civic engagement, and prescriptive visions for the "good city." To frame the debates and foreshadow the arguments of our authors, we take up each of these facets of urban theory in turn. In so doing, we draw attention to the differing perspectives—European, North American, South African—that are represented, but consider this issue in depth only as we conclude the chapter.

Representing the City

To a great extent, the turn to "representation" in urban theory is centered on urban form, with less attention given to the presentation of the social life in contemporary cities (but see King, 1996, and Lofland, 1998). The fascination with urban spatiality has been engendered not only by the physical city's solidity and visibility, its potential for immediate appropriation, but also by its surfacing of underlying economic, cultural, and governmental relationships. Thus, an insightful representation of the city's form constitutes a crucial step in achieving a useful representation of its dynamics.

Most writing on this theme draws on the imagery of either the industrial city of the United States or mid-20th-century European cities. This applies even when urban scholars write about Third World cities (Beavon, 1992). The Western dominance of urban morphology is well known and seems unavoidable.

The U.S. model is constructed around the industrial city of the late 19th and early 20th centuries, with its low-income neighborhoods woven into manufacturing districts and adjacent to commercial cores and its middle income neighborhoods beyond. Only in the late 19th century did the gentry move to the periphery, after World War II to be joined by a new professional and white-collar middle class. Out of this migration emerged the prototypical metropolis with its central city ringed by suburban enclaves. In the best cases, the commercial core became less dominant; in the worst cases, it (along with manufacturing districts) was hollowed out (Beauregard & Haila, 1997). The dynamics of development have been horizontal, with activities deconcentrated and decentralized from the center, a model of urban form strongly associated with the Chicago School of urban sociology.

By contrast, the European city is imagined as gathering a valuable public space around a core. There, residents meet and construct a common sense of the city. The complexity of overlapping powers—religious, political, cultural, economic—is conveyed by historical buildings next to one another and in the differences between adjacent neighborhoods. Expansion is radial and concentric. In Paris, Lyon, and Strasbourg, the historical center can easily be identified. Exceptions exist, of course; in Marseille, for example, the core has been hollowed out in the American fashion.

For urbanists in the United States, the European model is viewed with envy. Few of their cities can claim to be as vibrant during the day and evening and as desirable as places to live. Most U.S. cities, whether in the Rust Belt or the Sun Belt, are more prosperous on the periphery than in the center, with edge cities signaling the shift of corporate functions from the "traditional" central business district.

The Los Angeles depicted by Flusty and Dear and the cybercities revealed by M. Christine Boyer (Chapter 3 of this volume) are polycentric spaces that exacerbate the U.S. trend and make it increasingly difficult to match the European model, even in Europe. Consequently, urban theory is faced with a representational dilemma. Flusty and Dear argue that a change in urban form—with Los Angeles as emblematic—requires new metaphors. Boyer, on the other hand, begins with representation. Her position is that the computer has engendered a new sense of space that enables us to rethink how we portray the city even as that city is being physically transformed by advanced telecommunications (see also Mitchell, 1995).

Flusty and Dear propose a break with the modernism of the Chicago School. In its place, they offer a postmodern urbanism that better represents the disarticulated, multinucleated, polarized landscape of places like Los Angeles. For these two theorists, the city has become more unpredictable and less susceptible to formalization. But like the cities of modernity, postmodern cities still are subject to the forces of capitalism, even as capitalism itself has "morphed" into a new shape.

Any break from modernist urban theory, Flusty and Dear argue, requires a corresponding break from the language we use to portray the city; representation must respond to a changing reality. They propose new terms—*commudities* and *polyannarchy,* for example—that are meant to destabilize existing understandings and compel us to look differently at what are, for them, profoundly novel urban phenomena. The postmodern city is not the modern city, they argue. When we simply stretch prior terminology to cover new relationships and forms, we deny this sea change and distort understanding.

Boyer takes a different route to urban representation and form. For her, the motivating factor is the computer and the way in which it simultaneously makes us think differently about space even while transforming space through its power to overcome distances. The two-dimensional, fixed grid of the modern city is now replaced by the *n*-dimensional net of cyberspace in which the proximity of one node to another is merely happenstance. What is crucial are pathways (connectedness) and flows (Castells, 1989).

In this cyberspace, we confront the precarious nature of agglomeration, the ineffectiveness of boundaries (and their meanings), the disconnectedness of history, the reality of the fake, and the truth of the artificial. These dissolutions and distortions create new types of inequalities and exclusions. Space, however, remains significant. We still exist in places; our social nature resists the futuristic prognostications of cyberspace extremists. Boyer further understands that form is not easily separable from political function. She worries about the democratic city.

Despite the breathtaking leaps of Boyer and of Flusty and Dear from the comfortable confines of modernist urban theory, and despite their belief that a new city is here or soon to arrive, cities have yet to disengage fully from their history and the pull of gravity.

Cities formed during the Middle Ages still are the backbones of the current urban fabric in Europe. There, industrial urbanization has had a lesser impact in terms of segregation, for example, and the cycles of urbanization have been more homogeneous. The most "working class" neighborhoods in Paris today already were working class a century ago. By contrast, the American city tends to live in the present and project itself in the future (Baudrillard, 1986, brilliantly developed this theme). It avoids history; it constructs utopia outside of history.

Of course, this is a rhetorical generalization. New York, Boston, Chicago, or San Francisco would not fit this characterization. In these places, the tracks of prior cities still are visible through the collective weight of the past, a resistance of the agglomerated form and a resilience of the core, both perceived as the valued culture of American cities. They are sisters of European ones in this respect. As Paul Berman remarks, "What governs New York creativity is the strange tenacity of its ancient neighborhoods. . . . The determinism is geographical. Ghosts run the city" (Berman, 1988, pp. 32-33).

This European-U.S. contrast, however, has to be placed in a global context. Alan Mabin (Chapter 7 of this volume) correctly points out that it is (at root) a "north-south" comparison that draws from a common culture and identical historical trajectories through industrialization and urbanization. Particularly absent is the colonial experience (King, 1990) that had such a big impact on cities below the equator. Moreover, the tremendous expanse of poverty and destitution, not to mention the huge size of cities in "emerging" countries, has not been part of the "northern" experience, although in the "north" distinguishing globally between the "West" and the "East" is important. Given these material differences, we must ask (as Mabin also does) whether ways of representing cities across the globe can be universalized. To what extent do general terms fail us theoretically and practically?

Life in Johannesburg or Kigali (Rwanda) is much wilder than one finds in European or U.S. cities, and conditions are more precarious.

Theorists from the "north" have a tendency to celebrate a particular form of urbanity that is unlikely to be available in Soweto. They frequently note that well-off residents have not deserted Manhattan or the Loop and that Greenwich Village holds to a mini-culture of bohemianism much like Saint-Germain des Prés. The positive comparison stays within the Western culture, with non-Western forms of urbanity treated as undesirable or exotic. Normative assessments generate distinctions, and such value-laden distinctions make it difficult to represent, much less perceive, global commonalities.

Contested Terrains of Globalized Cities

Imperfect as it is, representation is the only way we have to convey the reality of cities, although even the solidity of urban form remains elusive. No more so than material conditions, representations are contested, and these contests become more complex as the urban world becomes further fragmented, polarized, and extended. Exacerbating these conditions is globalization, itself a chaotic constellation of alien dynamics.

The socioeconomic restructuring and the ensuing fragmentation produced by globalization, financial flux, and intercity competition constitute the deep background of the new urbanism whose representation is so contested. Residents of globalized cities are anxious about crime and dirt, about the security of their jobs, and about the ability of the economy (and its cities) to sustain unprecedented growth. Few in this affluent middle class, and almost none among political and economic elites, are committed to making decisions for the common good of future generations or of electoral minorities.

American cities, for example, are "better" *and* under full assault. They are more livable but still are losing their tax bases. Residents continue to follow delocalized jobs, while low-skilled migrants flock to the cities as the first step in their quest for prosperity. Cities in the United States and Europe, Asia and Africa, and South America and Australia engage each other in winner-take-all financial, economic, and demographic competition. Hierarchies of cities emerge with new centralities and new marginalities.

Saskia Sassen's pathbreaking book, *The Global City,* articulated the existence of new global cities anchored by financial services and corporate headquarter functions—contested terrains on a world scale

(Sassen, 1991). In elevating New York City, Tokyo, and London to the top of this particular hierarchy, she also paid attention to the local consequences that ensue as a city emerges as a node in a global network, specifically the implications for migration and low-wage labor. The linkages that make cities global are conduits for flows of people as well as capital and information, and the presence of new residents is equally a part of the disarticulation of citizenship to which Guido Martinotti (Chapter 8 of this volume) points from another direction.

In Chapter 5 of this volume, Saskia Sassen focuses on how global relations destabilize existing forms of political governance and corporate control. Specifically, she is concerned with the emergence of new claims—by immigrants who make different demands on the public realm and by transnational corporations that want nation- and city-states to relax certain regulations, denationalize their territories, and accommodate physical investments. The identities of people and places are in flux, and the processes that concentrate power and immigrants in these cities establish the basis for increased polarizations there. Global cities exhibit great wealth, conspicuously so, but they also engender a sizable working class of low-wage workers serving transnational corporations and their elites. New York City is the prime example, but neither London nor Tokyo can claim exclusion from this generalization.

If large European cities, such as Paris and Milan, can be seen as global, then it is in a soft version (Body-Gendrot, 1996). European cities and neighborhoods are indeed subjected to destructive economic pressures and to unemployment, poverty, crime, and violence. Yet, they seem to have more resources, specifically cultural ones, than do American cities for buffering social ills. The middle classes, less mobile than those in the United States, have not massively deserted working class neighborhoods in the central cities for sanctuaries in racially homogeneous suburbs. Unlike these situations, no neighborhood can claim a majority of minorities. Ethnic lines are blurred. The middle classes still are well represented in Belleville or in the Chinatown area of Paris (Rhein, 1998). Because more than one quarter of the Parisian labor force is in the public sector, during any workday there likely will be close contact between immigrant, homeless, and delinquent clients and those from the middle class who work as teachers, welfare administrators, police, judges, and social workers.

In the evening, in the mixed neighborhoods where they live, Parisians also might interact with a Tunisian grocer, an immigrant entrepreneur, an African street cleaner, or a young dealer selling in the street. Although more than one half of the French-dominant socioeconomic categories are concentrated in a few areas in the western and southern parts of Paris, the presence of a relatively significant number of employed foreigners in the same areas points to complex interconnections between immigration and urban restructuring. Still, important discrepancies in social conditions are experienced by both the dominant and dominated groups, and inequalities have been deepening due to the concentration of wealth in the top strata (Body-Gendrot, in press).

These foreigners are not simply of Paris or from Rabat; they are neither French nor Moroccan. Rather, and increasingly so, these are transnational groups living in multiple worlds, living on the borders of national meanings, practices, and spaces, as Michael Peter Smith reminds us in Chapter 6 of this volume. Their identities are complex, and their attachments to the city are tenuous, sometimes moving back and forth between their place of origin, many times forming ethnic enclaves within the city as havens against a foreign world, and often insisting on retaining prior cultural practices. Many of these immigrants (whether residing in European, Asian, American, or African cities) find themselves at the low end of labor queues, in low-rent housing markets, and faced with discrimination in numerous aspects of their lives.

The ways in which these transnational migrants deal with the conditions of the city set global cities in motion. Global cities, Michael Peter Smith argues, are not hard-wired into financial and corporate hierarchies (in a mild rebuke of Sassen) but rather are "soft," to use an earlier metaphor. To focus on their core functions is to misread them and to miss the ways in which globalization is manifested not in electronic transmissions but rather in social interactions that have a direct and immediate effect on those for whom uncertainty and risk have only a downside.

If the occupations of poorer groups and the places where the poor live are signs that the economy of European regional cities is transnational (as Michael Peter Smith claims), and if the polarization between the disadvantaged and the privileged of the "superior tertiary complex" (as formulated by Sassen) has increased, then the strategic role of the middle classes and the pressures they bring to bear on institutions as political actors, mediators, and interlocutors make European cities a case apart. Few distressed American cities can boast a dominant middle class, and fewer and fewer South African cities can do so. The growing precariousness felt by Europeans when jobs are scarce, real estate is too expensive, and the future for their children looms darkly ahead is openly and massively expressed in the streets and the media, in the polls and the ballot boxes, over and beyond generations and social circumstances. Yet, these resistances, in the form of social movements, are not absent in the United States and South Africa. In fact, they are quite robust.

Clearly, nation-states neither buffer cities from globalization (and might well do just the opposite) nor, because of their neoconservative tendencies (also manifest, yet softened by new labor regimes, in Europe), do they direct meaningful interventions at the new polarizations and injustices that are solidifying within cities. The outcome can be read not as a failure to make representative democracy manifest but as the normal functioning of the only democracy we are able to create.

Margit Mayer's comparative chapter on Germany and the United States (Chapter 10 of this volume) pulls us away from the ideals of democracy to the realities of social movements and to a different contestation of the city. From routinized cooperation with the state to opposition to competitive urban politics, contradictory and distinct voices resonate with the fragmented context of the postmodern city. Post-Fordist social movements unfold along different dimensions in response to the new international division of labor and the new global hierarchies having an impact on cities at the end of the 20th century. These movements produce a precarious democracy and have a problematic relation to citywide solidarities.

Implicit in Mayer's presentation is doubt about a local general interest, an implicit social cohesion. Instead, she sees the forces of disarticulation and injustice suppressing the democratic city yet calling forth social movements to counteract this tendency. Such movements do not always heed the global dimensions of local actions or the diversity of resources embedded in ethnicity, race, gender, and generations. No one will deny that centrifugal forces have a continuous impact on our cities, enmeshed as they are in global interdependencies. When the capacity of the state to structure the social realm weakens, as it has in Germany after the reunification and currently in France and Italy, for example, the local place appears as more dynamic than it was under a strong nation-state. Tensions and collective struggles are part of urban life, and one should not expect to find democratic reconciliation in the near future.

Of course, no social movement of the last half of the 20th century has been as important for democratic hopes as the anti-apartheid struggle in South Africa. Although it unfolded across the national landscape, the cities (both white cities and black townships) were the prime focal points. Apartheid itself was a profoundly spatial practice; political territory, natural landforms, man-made barriers, transportation systems, and legal boundaries were employed to segregate nonwhites from whites and yet enable nonwhite labor to support the economy. The city was, in one sense, "distorted" (Beauregard, 1994), a characterization that returns us to U.S. and European models. With the anti-apartheid victory and the election of Nelson Mandela as president in 1994, the political revolution came to a close, only to introduce the victors to new urban dynamics and the challenges of global capitalism.

Mabin, in his essay, notes the social and economic polarizations and fragmentations that have survived apartheid's defeat even as the post-apartheid city undergoes traumatic restructuring. Segregation persists, although at a finer scale and via more complex, less visible processes. The country's postcolonial legacy continues to fade, while global forces rush to occupy the expanding vacuum. The struggle is sustained, but without the revolutionary fervor of the liberation movement. How the city will change is unclear.

The conflicts that occur in cities are not confined to social movements. Daily resistances are sprinkled amid a multitude of irritations and small defeats that lead inexorably away from justice and democracy. Having once abandoned the city, Neil Smith (Chapter 9 of this volume) observes, the U.S. middle class now wants to recapture it and reassert control. This new revanchism involves a coalition of elected officials and place-based investors (mainly property developers) articulating their version of the good city, one mainly white and affluent and isolated from lingering urban ills—homelessness, street crime, prostitution, noise. The local government emphasizes "quality of life" and crime reduction. Developers reenergize and spread gentrification. Households riding the economic boom of the 1990s fuel the redevelopment transitions and, in return, are provided with opportunities to spend their money on luxury apartments, expensive restaurants, and upscale shopping, all to celebrate their "good" fortunes.

From this perspective, state policy is class policy. Yet, citizens of different countries have quite different expectations of the state, particularly in regard to its responsibilities for nurturing and ensuring social cohesion. In French cities, violent discontent erupts in the streets when the state does not fulfill that role. French, German, and Italian residents do not want less state. The current British prime minister, Tony Blair, has taken as his motto not to harm the middle classes, and he is not the only one. In the Old World, the state does not remain passive when market laws transform the economy and rend the social fabric. Comparing income groups over the past 10 years in New York City and Paris, sociologist Edmund Preteceille remarks that in the Parisian region, no massive pauperization is observed in the lowest income group, nor is absolute impoverishment of intermediate categories observed. "One cannot, therefore, characterize the impact of this globalization process as a dualization of the whole urban structure restoring the binary class cleavage of industrial capitalism in the city of the 19th century" (Preteceille, 1997, p. 106).

Nonetheless, and despite the differences across cities and countries, mayors and numerous elected officials at various levels between the global and the local do mount efforts to curb deindustrialization, urban poverty, social marginalization, and urban violence, although these efforts often are more defensive than quests for the ideals of solidarity. The urban issue has become a priority on their agendas as large cities have become the mirrors of transnationalism and the loci for change, heralding the major transformations about to overtake nation-states. Ignoring these changes is simply not an option, particularly given rampant urban competitiveness, the demands of highly mobile capital, and the indifference of nation-states.

A few mayors genuinely try to promote a social citizenship as a means of social prevention. Other mayors, as Neil Smith discusses, are less progressive, more mean-spirited toward immigrants and the poor, and intent on a totalizing view of the city. A majority of mayors are betwixt and between. They see themselves as committed to prosperity and improvement of the city's image, yet each pursues a different solidarity, with grave consequences for the inclusion and exclusion of different groups.

Cities are places where territorial subcultures and social groups are formed in specific and complex configurations. These configurations are maintained through what some social scientists call "urban governance," that is, the capacity to organize collective action, to form coalitions and partnerships toward specific goals (Bagnasco & LeGales, 1997). Policies, formal contracts, and partnerships bring diverse actors together to deliver a social package to treat populations and zones in distress.

Three forms of action are evident. One is the contract, limited in time and more often (but not always) involving businesses and local governments. The second is social movements that bubble up in opposition to shared rules, knowledge conventions, and common cultures. Third, there is the mobilization of heterogeneous networks turned toward the future. The modern city draws its dynamics from its capacity to crisscross such diversities, thereby mixing the self-interested frigidity of contracts with the reassuring warmth of shared cultural universes and the often cynical imagination of networks (Veltz, 1997, p. 64).

Social housing, neighborhood historical preservation, new means of public transportation, maintenance of public spaces, day care centers, and multi-service centers all are part of these efforts. As a caricature, Bagnasco and LeGales (1997) note that the European city can be characterized by "water power, *welfare,* and services on all floors" (p. 25, emphasis in original). However imperfect, European urban policies do attempt to slow the disintegrating social impact—"creative destruction," in Joseph Schumpeter's words (Schumpeter, 1942, pp. 81-86)—of economic restructuring.

Few urban commentators believe that the state and social movements can effectively resist the destructive forces of globalization and its attendant polarization. Mainly, they *hope* that this might be the case. Even the hope for state action, however, is dampened by the sense that states are moving further to the right and thereby putting more and more burden on cities to be competitive and compassionate, a difficult task. This awareness of neoconservatism's ascendancy, however, does not muffle the call for civic obligations to be met.

Under these conditions, democracy seems even more elusive, and contestation shifts from a pursuit of collective goals to a defensive localism (Weir, 1994) or a consumption-based individualism. Against the grain of such regressions, the city continues to offer the possibility, the hope, for democracy.

City of Democratic Hope

In the ideal democratic city, the walls have fallen. Across the divides of difference, people connect; they agree to differ. Collective memory is organized into a then and now that celebrates the present as a collective achievement. The vision is one of tolerance and diversity, shared values and complexity—not all for one, but the many for the all.

Where collective life and differences mutually coexist, democracy reigns. A full citizenship flourishes as the city's residents actively engage each other in a civil society that nurtures robust cultures and establishes the foundation for political engagement (Walzer, 1991). Socioeconomic polarities are minimized, and injustice, oppression, and exploitation are muffled. In this imagined city, frictions are not dispelled, failures are frequent, and disagreements are impassioned. The city of our imagination is not utopia.

For Thierry Paquot (Chapter 4 of this volume), this city is a place of friendship, of respect for others, and of acceptance of one's own strangeness. This city relies on surprise, tolerance, and diversity; it expects pluralism from its singularity. It asks its residents to "inhabit" the city rather than isolate themselves in "cellophane-wrapped worlds." It wants to last, Paquot writes; he wants it to last.

The "post-city," Paquot reminds us, incorporates the current planetary transformations not as a change in scale but rather as "something completely different." "A place exists only if it acts as a link, only if it is a place where something happens" (p. 81), he writes. From a distance, he evokes the American cities of the West, molded by freeways, as in a Joan Didion novel, with edge cities and gated communities sealed off like medieval castles. Residing there are people who also put a premium on difference; it leads them to the narrowest of togethernesses. Obvious for Paquot, however, is that the "urban" city is an encounter over which determinism has no hold.

Paquot also acknowledges the threats of disintegration weighing on the "urban" and killing the city. If the urban challenge bears fears and hopes, then the conclusion of his realistic essay bets on the plural, open, possible, and infinite for the city to come.

Such a city must emerge from the past. In the scenes of urbanity that Christian Ruby (Chapter 11 of this volume) unveils as so many vignettes, two tendencies are opposed, one promoting fear and the other promoting admiration. To the dual city, he contrasts a "city caught by its past, managed as if it were an ancient metropolis focused on the celebration of common abstract values and state experiences" (p. 241) whose sociability influences our appreciation of future involvements. He searches for a city based on relationships and anchored in the expansion of solidarity.

Europeans know that even if such a dream city comes to life, it does so only fleetingly and for an instant of grace. What matters is not the length but the seizure of people's imaginations. V-Day in 1945, May of 1968, and the celebration of France's 1998 World Cup victory revealed for a few intense moments that a city based on fraternity and generosity can take form and linger in collective memory. At such times, authorities denounce racism, privileges, and short-term vanities and propose goals for collective action. In response, people think about ideals and possibilities. Whose city is it? Ours! We: the blacks, Blancs, et Beurs. Few are those who would dare reject urbanity in such fully urban moments.

Ruby's metaphor of the archipelago nicely captures this notion of separate but together or, as Young (1996) puts it, "together in difference." In the urban archipelago, "common actions are constantly rebuilt" amid responsibility and controversy; these moments of urbanity still are only moments. The daily life of the city neither contains nor can sustain them on an ongoing basis. The city of democratic hopes must weather the rarity of peak collective experiences. We keep looking toward the future for an ideal that never moves closer.

To search for the city of the future, Martinotti warns in Chapter 8, is to forget that we already live there. What is challenging—a premise that all of our authors acknowledge—is the intuition that new stakes, realignments, and cleavages are generated by the profound transformations that cities undergo. These transformations occur all around us, but we are unable to fathom their full consequences. Consequently, we express them as ideologically and culturally loaded predispositions. At worst, we slip into the incurably romantic delusion that past arrangements can be preserved.

Martinotti elaborates on Ruby's clear-sighted yet fragile vision of the city. He notes that spaces for democracy have to be constantly negotiated and renegotiated in the face of ever encroaching privatized and commercialized realms. The public space, symbolized by a density of contacts, becomes a symbol for citizenship rights. In the *civitas,* collective images abound—churches, common squares, plazas, all the products of a holistic principle of social space that represents the city in its entirety.

Martinotti views these public spaces and the democratic city that embraces and nurtures them as threatened by the deepening divisions of social and spatial activity, an extension of the famed spatial distancing of home from workplace. This latter separation, he insists, was only one stage in the evolution of the metropolis. The city now experiences not just residents and commuters but also users (those who come to shop, be entertained, and act the role of tourists) and businesspeople who come from around the globe. Many of the users and nearly all of the businesspeople do not reside nearby. When they leave the city, they escape its shadow and its obligations. Their connections to the city are not tenuous but rather are easily replaced by connections to other cities. The expansion of nonresident populations calls into question the meaning of urban citizenship and the functioning of the political realm.

The plea for the city as a unifying political space also is defended by Richard Sennett (Chapter 13 of this volume). He expresses a nostalgia for cities of the past—Paris, Jerusalem, Athens. He argues that the design of both the agora and the Greek theater (*pnyx*) were meaningful symbols and functional spaces to which all citizens could relate. The agora fostered an awareness of differences. In the pnyx, citizens engaged in debate and decision making, with their attention concentrated on the orator. In these centralized sites, they witnessed the workings of government. Sennett is aware, however, that democracy comes with a cost—the realistic acceptance of fragmentation, complexity, and uncertainty. One is overwhelmed by powerlessness in a theater that gradually loses its unity of place under the forces of growing global interdependence. Differences lead to isolation, and in that state, we no longer engage with each other. The ideals of social justice and cohesion similarly recede from our grasp. Sennett, too, believes in political virtues and the importance, if not the sufficiency, of democratic spaces.

Susan S. Fainstein (Chapter 12 of this volume) takes up the theme of democracy. To the value of participation, she adds justice, equality, diversity, and sustainability. She explores the "ideal of a revitalized, cosmopolitan, just, and democratic city" (p. 250) under the premise that its values are "beneficent yet discordant" (p. 252) and that its realization requires that we identify the sources of social injustice and act to dampen them. Consequently, we are faced with the possibility that these values and principles cannot coexist or are not compatible. In true planning fashion, we are forced to make trade-offs; we are compelled to live in a world where interests and identities preclude attainment of the urban imaginary.

In an attempt to salvage hope from possible despair, Fainstein directs our attention to Amsterdam. She offers a European city as "the best available model" (p. 262) of the just, democratic, cosmopolitan and sustainable city. Her praise is tempered by her practicality; the conditions that make Amsterdam such a livable city are likely too culturally, historically, and place specific to be exportable. Her implication is that we are left, for the most part, with places like New York City, where even in the midst of an economic boom, inequalities persist and harden, racial tensions abound, and public services and infrastructure are neglected. Tolerance can tolerate only so much diversity; limits exist. New

York City is closer to the global norm than is Amsterdam. Nonetheless, Fainstein persists. She advises a search for a common ground and the launching of counterinstitutions that will (hopefully) direct cities toward the values that she and other urbanists espouse.

The message of these essays is that cities remain places of ideological and cultural resistance, and they offer hope of empowerment and innovation. The seeds of a democratic city continue to reproduce in historical memory. Urban theorists, plying their trade as social critics, point to the possibilities. Their claims are normative, and unapologetically so.

Conclusions

To end by stating that urban theory today is in delightful disarray would be to imply more disagreement than actually exists. Large consensual patches abut vast areas of uncertainty, with the usual defenders of debunked perspectives and eccentric formulations scattered about the fringes. Urban theorists, despite Boris Pasternak's claim that "gregariousness is always the refuge of mediocrities" (Pasternak, 1958/1991, p. 9), are disinclined to seek the splendor of intellectual isolation.

For the most part, urban theorists agree that something new is happening or has happened to contemporary cities. Globalization is one of the culprits, and it spreads this "happening" to all points of the compass. They also agree that new modes of representation are needed. This might simply mean the adjustment of existing concepts or might extend to the introduction of novel metaphors and unusual frameworks. A third area of agreement involves the deepening polarization and disarticulation of cities; few claim an irresistible trend to wholeness and consistency, although many witness pockets of urbanity.

Less widely embraced is the move to more explicit normative models of the city (Sayer & Storper, 1997). This has numerous roots—the melding of political theory with urban issues (particularly around the themes of citizenship and deliberative democracy), the related concern with injustices and inequalities, a dissatisfaction with the withdrawal of nation-states from urban policy, the slippage from quantitative analysis to cultural musings, and the slow seepage of public philosophy into the social sciences. The normative turn is well represented in the chapters of this book, and it is not confined to Europeans, North Americans, or South Africans. It always has been part of an urban theory whose modern roots lie in social reform. Only now is it being made explicit.

Other issues attract more ambivalence—the importance of social movements, the potential of a progressive state to emerge and function compassionately, and the extent to which citizenship disengaged from workplace struggles is the correct path to a just and democratic city. The allure of cyberspace also triggers equivocation. It appears both in the celebration of telecommunications technology by global theorists and in the use of cyberspace technology and metaphors to rethink the tangibility and meaning of space.

Urban theorists bow to the representational, although fewer of them—Flusty and Dear are the exceptions in this volume—embrace its disorienting principles. The validity of multiple points of view also is widely asserted even as most theorists prefer one perspective over others, and repeatedly so. We know of few urban scholars who would have the patience of a Kublai Khan to sit through Marco Polo's evocations of the city, much less adopt this as their mode of inquiry. Multiple perspectives debilitate a social science meant to explain and clarify.

As cities become more multicultural, however, we are pushed back to the challenge of multiple perspectives and representations. Within any given (global?) city, the north and the south coexist and traces of Europe and Africa, the United States and Central America, and Japan and Vietnam surface. European cities are different from cities in South Africa, Indonesia, or Australia. What is a theorist, whose imagination has been nurtured by a specific culture and whose career is (more rather than less) bound by it, to do? Even with the best of intentions, can we occupy multiple global positions? On the one hand, as Fish (1980) quips, "Relativism is a position one can entertain; it is not a position one can occupy" (p. 319). On the other hand, a global perspective proposes a universal point of view that we might wish to reject.

Divergent viewpoints between those whose minds extract the city as an idealized potential and those who stick closely to the struggles and conditions that give those ideals value are unavoidable. Philosophers and social scientists, idealists and materialists, are a few of the simple distinctions. More germane is that scholarly traditions are represented here, with the European academics drawing on their positive urban experiences and the greater national commitment to cities and their U.S. and South African counterparts more attuned to divisions and animosities and the ravages of racism and capitalism. The distinction, however, is not sharp. Democratic hope is not hemmed in by national boundaries or cultural origins but rather is widely held.

A judge or prosecutor does not behave in the same way in Dallas and Brooklyn, in Paris and Nice, in London and Liverpool. Expectations, pressures, and collective cultures have a lingering effect on individual behaviors. Apart from non-places, of which Gertrude Stein has been oft-quoted as saying "There is no there there," many cities (but fewer and fewer) have identifiable and instantaneous identities. This is what emerges from city effects. The historical evolution of civic cultures and their repertoires mold the ways in which they deal with the problems of fractures and social deviance generated by global forces.

To the extent that place is significant for intellectual work (i.e., to the extent that the flow of ideas across the globe still requires their production by individuals with place-specific histories), three historical, sociospatial perspectives are represented in this book: Western European, U.S., and South African. Because ideas (and intellectuals) travel across national boundaries and because knowledge is increasingly produced in multiple locations, however, these geographic and cultural labels elaborate rather than determine the arguments that are made. At one extreme are "free-floating" intellectuals unfettered by history and geography; at the other are those who cannot escape (or who do not even recognize) the influence of time and place (Boggs, 1993; Walzer, 1988). The first is impossible to realize, and the second is impossible to escape. In between fall the authors in this book.

In taking up this theme—the historical-geographical specificity of urban theorizing—we embark on perilous ground, risking immersion in the quicksand of spatial and cultural determinisms. Yet, it seems to us that writers of European origins, at least those included here, tend to romanticize a consensual utopia, much like that formulated by the symbolic production of Augustinus in *The City of Angels.* American authors, broadly cast and not without exception, are more likely to denounce the pernicious influence that capitalism has on the city, acknowledging hierarchical differences produced by investment logics that are visible in the segregated spaces of the city itself. Moreover, they often avoid broad normative theorizing, although that is changing, as Fainstein attests in her essay.

To give one example, South African urban theorists, having faced decades of apartheid, understand the need for reconciliation and the pressing challenges of turning ultra-segregated cities into multi-ethnic, democratic spaces (Parnell, 1997). As aware as Americans of the forces of disarticulation, and inspired by the movement's political success, they hope for a just city despite the immense economic inequalities and racial tensions that persist. European cities have, by comparison, been less feral. Europeans are more likely to consider their cities to be national treasures, whereas Americans are more likely to abandon theirs at a moment's notice or to treat them as commodities or investment options whose value is, at best, fleeting. South Africans know that their future lies in the cities.

Until the advent of national societies during the 19th century, cities molded collective imagination and social life. In the New World, an anti-urban culture distanced elites from cities and left the cities as only one element in the urban landscape (Beauregard, 1993, pp. 9-26). Contemporary American theorists, although proclaiming their love of cities, nevertheless mirror this attitude in their obsession with cities as contaminated by capitalist forces that limit the margins for action of public authorities and their "private" business partners. In Europe, as admirably demonstrated by Weber (1921/1982), the city from the beginning formed a local and political society; it has continued to function as a locus where interest groups are amalgamated and represented. With the difficulties currently experienced by nation-states and the increasingly transnational nature of modern society, this urban imaginary has returned to challenge nation-based citizenship and imaginations.

The city no longer is the enclosed space it was during the Middle Ages or during the early settlement of the United States. Still, its political significance remains formidable, not as much for the capacity to govern (higher in Europe than in the United States or South Africa) as for the concentration of potential to engender democratic practices and embrace a political citizenship embedded in civil society.

We are witnessing a turning point in urban theory. Everywhere the lack of new ideas in the analysis of the city has been part of the urban crisis. The future of the democratic city is fragile indeed; the challenge of an uncertain future comes from the fluidity, discontinuity, and uncertainty that we have not yet learned to master. Although we do not know what directions eventually will prove fruitful or what destinations will be achieved, we nonetheless offer this book as both a documentation of that search and a dare to those who still are dissatisfied.

REFERENCES

Bagnasco, A., & LeGales, P. (Eds.). (1997). *Villes en Europe.* Paris: L'Harmattan.

Baudrillard, J. (1986). *Amerique.* Paris: Grasser.

Beauregard, R. A. (1993). *Voices of decline: The postwar fate of U.S. cities.* Oxford, UK: Blackwell.

Beauregard, R. A. (1994). *Distorted cities* [pamphlet]. Harvey S. Perloff Lecture (May 5), Graduate School of Architecture and Urban Planning, University of California, Los Angeles.

Beauregard, R. A., & Haila, A. (1997). The unavoidable incompleteness of the city. *American Behavioral Scientist, 41,* 327-341.

Beavon, K. (1992). The post-apartheid city. In D. M. Smith (Ed.), *The apartheid city and beyond* (pp. 231-242). London: Routledge.

Berman, M. (1982). *All that is solid melts into air.* New York: Simon & Schuster.

Berman, P. (1988, March 15). Mysteries and majesties of New York. *The Village Voice,* pp. 32-33.

Body-Gendrot, S. (1996). Paris: A soft global city? *New Community, 22,* 595-606.

Body-Gendrot, S. (in press). *The social control of cities.* Oxford: Blackwell.

Boggs, C. (1993). *Intellectuals and the crisis of modernity.* Albany: State University of New York Press.

Bohman, J. (1996). *Public deliberation.* Cambridge, MA: MIT Press.

Calvino, I. (1974). *Invisible cities.* New York: Harcourt, Brace.

Castells, M. (1989). *The informational city.* Oxford, UK: Blackwell.

Fish, S. (1980). *Is there a text in this class?* Cambridge, MA: Harvard University Press.

King, A. (1990). *Global cities.* London: Routledge.

King, A. (Ed.). (1996). *Re-presenting the city.* New York: New York University Press.

Lofland, L. H. (1998). *The public realm.* New York: Aldine de Gruyter.

Merrifield, A., & Swyngedouw, E. (Eds.). (1997). *The urbanization of injustice.* New York: New York University Press.

Mitchell, W. J. (1995). *City of bits.* Cambridge, MA: MIT Press.

Parnell, S. (1997). South African cities: Perspectives from the ivory tower of urban studies. *Urban Studies, 34,* 891-906.

Pasternak, B. (1991). *Doctor Zhivago.* New York: Pantheon. (Originally published in 1958)

Preteceille, E. (1997). Segregation, classes et politique dans la grande ville. In A. Bagnasco & P. LeGales (Eds.), *Villes en Europe* (pp. 99-128). Paris: L'Harmattan.

Rhein, C. (1998). Globalization, social change, and minorities in metropolitan Paris: The emergence of new class patterns. *Urban Studies, 35, 429-447.*

Sassen, S. (1991). *The global city.* Princeton, NJ: Princeton University Press.

Sayer, A., & Storper, M. (1997). Ethics unbound: For a normative turn in social theory. *Environment & Planning D, 15,* 1-17.

Schumpeter, J. A. (1942). *Capitalism, socialism, and democracy.* New York: Harper & Row.

Toulmin, S. (1990). *Cosmopolis.* Chicago: University of Chicago Press.

Veltz, P. (1997). Les villes europêenes dans l'économie mondiale. In A. Bagnasco & P. LeGales (Eds.), *Villes en Europe* (pp. 47-66). Paris: L'Harmattan.

Walzer, M. (1988). *The company of critics.* New York: Basic Books.

Walzer, M. (1991). The idea of civil society. *Dissent, 38,* 293-304.

Weber, M. (1982). *La ville.* Paris: Aubier-Montaigne. (Originally published in 1921)

Weir, M. (1994). Urban poverty and defensive localism. *Dissent, 41,* 337-342.

Young, I. M. (1990). *Justice and the politics of difference.* Princeton, NJ: Princeton University Press.

Young, I. M. (1996). Together in difference. In W. Kymlicka (Ed.), *The rights of minority cultures* (pp. 155-179). New York: Oxford University Press.

Part I

Dilemmas of Representation

2

Invitation to a Postmodern Urbanism

STEVEN FLUSTY
and MICHAEL DEAR

The identification of historical periods is contested terrain, with the most heated contention focused on when (or indeed whether) a break inaugurating a new epoch at the expense of the old has actually occurred. Such contests can become particularly bloody when the new epoch is declared contemporaneously with its manifestation in quotidian political, economic, and sociocultural life. This certainly has been the case in efforts to proclaim, or deny, a radical break indicative of a transition from a modern to a postmodern condition.

No clear consensus exists about the nature of this ostensible break. Some analysts have declared the current condition to be nothing more than business as usual, only faster—not postmodern at all but rather a hypermodern or supermodern phase of advanced capitalism.[1] Others have noted that, in and of itself, the accelerated pace of change in all aspects of our global society is sufficient for us to begin to speak of revolution. Critical to negotiating such debates, however, is the recognition that periodization is a procrustean activity, frequently suppressing both variations within epochs and continuities between them so as to facilitate the application of contemporary categorizations. As a result, periodizing breaks might well be less indicative of actual disjunctions in the continuum of events themselves than of the moment at which one is forced by circumstances, kicking and screaming if need

AUTHORS' NOTE: This chapter was first published in the *Annals of the Association of American Geographers* (1998), Volume 88, Number 1 (pp. 50-72). This revised version appears with the kind permission of its publisher. We also are indebted to Robert Beauregard for his thorough and detailed editorial attention.

be, to acknowledge that one's preconceived formulas for apprehending the world no longer are tenable in light of changes in the world itself.

Thus, this chapter proceeds from an invocation of Jacques Derrida, who invited those interested in assessing the extent and volume of contemporary change to "rehearse the break," intimating that only by assuming that a radical break had occurred would we become capable of recognizing that the ground beneath our feet already has shifted. Similar advice was offered by C. Wright Mills. Mills (1959) believed that it was vital to conceptualize the categories of change so as to "grasp the outline of the new epoch we suppose ourselves to be entering" (p. 166).

Have we arrived at a radical break in the way in which cities are developing, demanding a similarly radical break from old models of urbanism that are increasingly ill equipped to address contemporary urban experience? Is there something called a *postmodern urbanism,* and if so, can we begin to derive templates that define its critical dimensions?[2]

Our inquiry is based on a simple premise: Just as the central tenets of modernist thought have been undermined, its core evacuated and replaced by a rush of competing epistemologies, so too have the traditional logics of earlier urbanisms evaporated. In the absence of a single new imperative, multiple urban (ir)rationalities are competing to fill the void. The concretization and localization of these effects, global in scope but generated and manifested locally, are creating the geographies of postmodern society—a new time-space fabric.[3]

■ Ways of Seeing: Southern Californian Urbanisms

Theories of urban spatial structure for the most part have been haunted by, if not thoroughly grounded in, the organic metaphors of the Chicago School of urbanism. The zonal or concentric ring theory, the ecological model of invasion-succession-segregation, and the later modifications of the multiple nulclei model are well known. Recently, this Chicago School vision has been challenged and perhaps displaced by Los Angeles–based theoreticians who assert that Southern California is a suggestive prototype—a polyglot, polycentric, and polycultural pastische that is engaged in the redrawing of American cities. These scholars' investigations lay the groundwork for a postmodern urbanism that ultimately might restructure urban theory itself. Central to this project has been the repositioning of Los Angeles from the role of exception to that of exemplar.

Most world cities have instantly identifiable signatures; think of the boulevards of Paris, the skyscrapers of New York, or the churches of Rome. Los Angeles long appeared as a city without a common narrative except perhaps the freeways or the more generic iconography of the bizarre. A quarter of a century ago, Rayner Banham provided an enduring map of the Los Angeles landscape. To this day, it remains powerful, evocative, and instantly recognizable. Banham (1973) identifies four basic ecologies: *surfurbia* (the beach cities: "The beaches are what other metropolises should envy in Los Angeles. . . . Los Angeles is the greatest city-on-the-shore in the world" [p. 37]), *the foothills* (the privileged enclaves of Beverly Hills, Bel Air, etc., where the financial and topographical contours correspond almost exactly), *the plains of Id* (the central flatlands: "an endless plain endlessly gridded with endless streets, peppered endlessly with ticky-tacky houses clustered in indistinguishable neighborhoods, slashed across by endless freeways that have destroyed any community spirit that may have once existed, and so on . . . endlessly" [p. 161]), and *autopia* ("[The] freeway system in its totality is now a single comprehensible place, a coherent state of mind, a complete way of life" [p. 213]).

For Douglas Suisman, it is not the freeways but rather the boulevards that determine the city's overall physical structure. A boulevard is a surface street that "(1) makes arterial connections on a metropolitan scale, (2) provides a framework for civic and commercial destination, and (3) acts as a filter to adjacent residential neighborhoods." He argues that boulevards do more than establish an organizational pattern; they constitute "the irreducible armature of the city's *public space*" and are charged with social and political significance that cannot be ignored. Usually sited along the edges of former *ranchos,* these vertebral connectors today form an integral link among the region's municipalities (Suisman, 1989, pp. 6-7, emphasis in original).

For Edward Soja, Los Angeles is a decentered, decentralized metropolis powered by the insistent fragmentation of post-Fordism, that is, an increasingly flexible, disorganized regime of capitalist accumulation. Accompanying this shift is a postmodern consciousness, a cultural and ideological reconfiguration altering how we experience social being. The center holds, however, because it functions as the urban panopticon, that is, the strategic surveillance point for the state's exercise of social control. Out from the center extends a melange of "wedges" and "citadels," interspersed between corridors formed by the boulevards. The consequent urban structure is a complicated quilt, fragmented yet bound to an underlying economic rationality: "With exquisite irony, contemporary Los Angeles has come to resemble more than ever before a gigantic agglomeration of theme parks, a lifespace composed of Disneyworlds" (Soja, 1989, p. 246).

These three sketches provide differing insights into the Los Angeles landscapes. Banham (1973) considers the city's overall torso and identifies three basic components (surfurbia, plains, and foothills) as well as connecting arteries (freeways). Suisman (1989) shifts our gaze away from principal arteries to the veins (the boulevards) that channel everyday life. Soja (1989) considers the body-in-context, articulating the links between political economy and postmodern culture to explain fragmentation and social differentiation. All three writers maintain a studied detachment from the city, as though a voyeuristic, top-down perspective is needed to discover the rationality inherent in the cityscape.

A postmodern sensibility would relinquish the modernism inherent in such detached representations of the urban text. What would a postmodernism from below reveal?

One of the most prescient visions anticipating a postmodern cognitive mapping of the urban is Jonathan Raban's *Soft City,* a reading of London's cityscapes. Raban divides the city into hard and soft elements (Raban, 1974). The former refers to the material fabric of the built environment—the streets and buildings that frame the lives of city dwellers. The latter, by contrast, is an individualized interpretation of the city, a perceptual orientation created in the mind of every urbanite.[4] The relationship between the two is complex and even indeterminate. Raban makes no claims to a postmodern consciousness, yet his invocation of the relationship between the cognitive and the real leads to insights that are unmistakably postmodern in their sensitivities.

Ted Relph, one of the first geographers to catalog the built forms of postmodernity, describes postmodern urbanism as a self-conscious and selective revival of elements of older styles, although he cautions that postmodernism is not simply a style but also a frame of mind (Relph, 1987, p. 213). He observes how the confluence of many trends—gentrification, heritage conservation, architectural fashion, urban design, and participatory planning—caused the collapse of the modernist vision of a future city filled with skyscrapers and other austere icons of scientific rationalism. The new urbanism is principally distinguishable from the old by its *eclecticism.*

Relph's (1987) periodization of 20th-century urbanism involves a premodern transitional period (up to 1940), an era of modernist cityscapes (after 1945), and a period of postmodern townscapes (since 1970). The distinction between cityscape and townscape is crucial to his diagnosis. Modernist cityscapes, he claims, are characterized by five elements:

1. Megastructural bigness (few street entrances to buildings, little architectural detailing);
2. Straight-space/prairie-space (city center canyons, endless suburban vistas);
3. Rational order and flexibility (the landscapes of total order, verging on boredom);
4. Hardness and opacity (freeways, the displacement of nature); and
5. Discontinuous serial vision (deriving from the dominance of the automobile). (pp. 242-250)

Conversely, postmodern townscapes are more detailed, handcrafted, and intricate. They celebrate difference, polyculturalism, variety, and stylishness. Their elements are as follows:

6. Quaintspace (a deliberate cuteness);
7. Textured facades (for pedestrians, rich in detail, often with an "aged" appearance);
8. Stylishness (appealing to the fashionable, chic, and affluent);
9. Reconnection with the local (involving deliberate historical/geographical reconstruction); and
10. Pedestrian-automobile split (to redress the modernist bias toward the car). (pp. 252-258)

Raban's (1974) emphasis on the cognitive and Relph's (1987) emphasis on the concrete underscore the importance of both dimensions in understanding sociospatial urban processes. The palette of urbanisms that arises from merging the two is thick and multidimensional. We construct that palette (what we earlier described as a template) by examining empirical evidence of recent urban developments in Southern California. We take our lead from what exists and then move quickly to a synthesis that prefigures a proto-postmodern urbanism.

Edge Cities

Joel Garreau notes the central significance of Los Angeles in understanding contemporary metropolitan growth in the United States. He asserts that "every single American city that *is* growing, is growing in the fashion of Los Angeles" and refers to Los Angeles as the "great-grandaddy" of edge cities (Garreau, 1991, p. 3, emphasis in original). For Garreau, edge cities represent the crucible of America's urban future.

The classic location for contemporary edge cities is at the intersection of an urban beltway and a hub-and-spoke lateral road. The central conditions that have propelled such development are the dominance of the automobile and the associated need for parking, the communications revolution, and the entry of women into the labor market in large numbers. Garreau (1991) identifies three basic types of edge city: "uptowns" (peripheral pre-automobile settlements that subsequently have been absorbed by urban sprawl), "boomers" (the classic edge cities, located at freeway intersections), and "greenfields" (the current state of the art, "occurring at the intersection of several thousand acres of farmland and one developer's monumental ego") (p. 116).

An essential feature of the edge city is that politics is not yet established there. Into the political vacuum moves a shadow government—a privatized proto-government that is essentially a plutocratic alternative to normal politics. Shadow governments can tax, legislate for, and police their communities, but they rarely are accountable, being responsive primarily to wealth (as opposed to numbers of voters) and subject to few constitutional constraints (Garreau, 1991, p. 187). Jennifer Wolch describes the rise of the shadow state as part of a societywide trend toward privatization. In edge cities, community is scarce, and the walls that typically surround such neighborhoods are social boundaries that act as community "recognizers," not community "organizers" (Wolch, 1990, pp. 275-281). During the edge city era, Garreau (1991) notes, the term "master-planned community" is little more than a marketing device (p. 301).[5]

Privatopia

Privatopia, perhaps the quintessential edge city residential form, is private housing based in common interest developments (CIDs) and administered by homeowners associations. Sustained by an expanding catalog of covenants, conditions, and restrictions (or CC&Rs, the proscriptive constitutions formalizing CID behavioral and aesthetic norms), privatopia has been fueled by a large dose of privatization and been promoted by an ideology of "hostile privatism" (McKenzie, 1994, p. 19). It has provoked a culture of nonparticipation.

Evan McKenzie warns that, far from being a benign or inconsequential trend, CIDs already define a new norm for the mass production of housing in the United States. McKenzie (1994) notes how this "secession of the successful" (the phrase is Robert Reich's) has altered concepts of citizenship in which "one's duties consist of satisfying one's obligations to private property" (p. 196).

In her futuristic novel of Los Angeles wars between walled-community dwellers and those beyond the walls, *Parable of the Sower,* Octavia Butler envisions a dystopian, privatopian future. It includes a balkanized nation of defended neighborhoods at odds with one another, where entire communities are wiped out for a handful of fresh lemons or a few cups of potable water and where company-town slavery is attractive to those who are fortunate enough to sell their services to the hyper-defended enclaves of the very rich (Butler, 1993).

Cultures of Heteropolis

One of the most prominent sociocultural tendencies in contemporary Southern California is the rise of minority populations (Ong, Bonacich, & Cheng, 1994; Roseman, Laux, & Thieme, 1996; Waldinger & Bozorgmehr, 1996). Provoked to comprehend the causes and implications of the 1992 civil disturbances in Los Angeles, Charles Jencks zeroed in on the city's diversity as the key to the city's emergent urbanism: "Los Angeles is a combination of enclaves with high identity and multi-enclaves with mixed identity, and taken as a whole, it is perhaps the most heterogeneous city in the world" (Jencks, 1993, p. 32). Such ethnic pluralism has given rise to what Jencks calls a hetero-architecture. In it, "there is a great virtue, and pleasure, to be had in mixing categories, transgressing boundaries, inverting customs, and adopting the marginal usage" (p. 123). The vigor and imagination underlying these intense cultural dynamics are everywhere evident in the region, from the diversity of ethnic adaptations (Park, 1996), to the concentration of cultural producers (Molotch, 1996), to the hybrid complexities of emerging cultural forms (Boyd, 1996, 1997).

Jencks (1993) views hetero-architecture as a hopeful sign, the main point of which "is to accept the different voices that create a city, suppress none of them, and make from their interaction some kind of greater dialogue" (p. 75). This is especially important in a city where *minoritization,* "the typical postmodern phenomenon where most of the population forms the 'other,' " is the order of the day and where most city dwellers feel distanced from the power structure (p. 84). Despite Jencks's optimism, other analysts have observed that the same Southern California heteropolis has to contend with more than its share of socioeconomic polarization, racism, inequality, homelessness, and social unrest (Anderson, 1996; Baldassare, 1994; Bullard, Grigsby, & Lee, 1994; Gooding-Williams, 1993; Rocco, 1996; Wolch & Dear, 1993). These characteristics are part of a sociocultural dynamic that is provoking the search for innovative solutions in labor and community organizing (Pulido, 1996) as well as in interethnic relations (Abelmann & Lie, 1995; Martinez, 1992; Yoon, 1997).

City as Theme Park

California in general and Los Angeles in particular often have been promoted as places where the "American (suburban) Dream" is most easily realized. Converted to built form, however, such dreams readily become marketable commodities, that is, salable prepackaged landscapes engineered to satisfy fantasies of suburban living.[6]

Many writers have used the "theme park" metaphor to describe the emergence of such variegated cityscapes. Michael Sorkin, in a collection of essays appropriately titled *Variations on a Theme Park,* describes theme parks as places of simulation without end, characterized by aspatiality plus technological and physical surveillance and control. The precedents for this model can be traced back to the World's Fairs, but Sorkin (1992) insists that something wholly new is now emerging: "The 800 telephone number and the piece of plastic have made time and space obsolete," and these instruments of "artificial adjacency" have eviscerated the traditional politics of propinquity (p. xi). Sorkin observes that the social order always has been legible in urban form; for example, traditional cities have adjudicated conflicts via the relations of public places such as the agora or piazza. In today's "recombinant city," however, conventional legibilities have been obscured and/or deliberately mutilated. The phone and modem have rendered the street irrelevant. The new city threatens an "unimagined sameness" characterized by the loosening of ties to any specific space; rising levels of surveillance, manipulation, and segregation; and theme park elements. What is missing in this new cybernetic suburbia is not a particular building or place but rather the spaces between, that is, the connections that make sense of forms. What is missing is connectivity and community.

In extremis, California dreamscapes become simulacra. Soja (1992) identifies Orange County as a massive simulation of what a city should be. He describes it as "a structural fake, an enormous advertisement, yet functionally the finest multipurpose facility of its kind in the country." Calling this assemblage *exopolis,* or the city without, Soja asserts that "something new is being born here," based on the hyperrealities of more conventional theme parks such as Disneyland (p. 101). The exopolis is a simulacrum, an exact copy of an original that never existed, within which image and reality are spectacularly confused. In this "politically numbed" society, conventional politics is dysfunctional. Orange County has become a "scamscape," notable principally as home of massive mail fraud operations, savings and loan failures, and county government bankruptcy (p. 120).

Fortified City

The downside of the Southern Californian dream has been the subject of countless dystopian visions in histories, movies, and novels.[7] In one powerful account, Mike Davis notes how Southern Californians' obsession with security has transformed the region into a fortress. This shift is accurately manifested in the physical form of the city, which is divided into fortified cells of affluence and places of terror where police battle the criminalized poor. These urban phenomena, according to Davis, have placed Los Angeles "on the hard edge of postmodernity" (Davis, 1992b, p. 155). The dynamics of fortification involve the omnipresent application of high-tech policing methods to the "high-rent security of gated residential developments" and "panopticon malls." It extends to "space policing" including a proposed satellite observation capacity that would create an invisible Haussmannization of Los Angeles. In the consequent "carceral city," the working poor and destitute are spatially sequestered on the "mean streets" and excluded from the affluent "forbidden cities" through "security by design."

Elaborating on Davis's (1992b) fortress urbanism, Steven Flusty observes how various types of fortification have extended a canopy of suppression and surveillance across the entire city. His taxonomy of interdictory spaces identifies how spaces are designed to exclude by function and cognitive sensibilities. Some spaces are passively aggressive; space concealed by intervening objects or grade changes is "stealthy," and spaces that may be reached only by means of interrupted or obfuscated approaches is "slippery." Other spatial configurations are more assertively confrontational—deliberately obstructed "crusty" space demarcated by walls and checkpoints, inhospitable "prickly" spaces featuring unsittable benches in areas devoid of shade, and "jittery" space ostentatiously saturated with surveillance devices (Flusty, 1994, pp.16-17).

Flusty (1994) notes how combinations of interdictory spaces are being introduced "into every facet of the urban environment, generating distinctly unfriendly mutant typologies" (p. 20). Some are indicative of the pervasive infiltration of fear into the home including the bunker-style "blockhome," affluent palisaded "luxury laager" communities, and low-income residential areas converted into "pocket ghettos" by military-style occupation. Other typological forms betray a fear of the public realm, as with the fortification of commercial facilities into "strongpoints of sale" and the self-contained "world citadel" clusters of defensible office towers.

One consequence of the sociospatial differentiation described by Davis (1992b) and Flusty (1994) is an acute fragmentation of the urban landscape. Commentators who remark on the strict division of residential neighborhoods along race and class lines miss the fact that the Los Angeles microgeography is incredibly volatile and varied. In many neighborhoods, simply turning a street corner will lead the pedestrian/driver into totally different social and physical configurations.[8]

Post-Fordist Technopoles

Many observers agree that one of the most important underlying shifts in the contemporary political economy is from a Fordist to a post-Fordist industrial organization. In a series of important books, Scott and Storper have portrayed the burgeoning urbanism of Southern California as a consequence of this deep-seated structural change in the capitalist political economy (Scott, 1988a, 1988b, 1993; Storper & Walker, 1989). Scott's basic argument is that there have been two major phases of urbanization in the United States. The first related to an era of Fordist mass production during which the paradigmatic cities of industrial capitalism (e.g., Detroit, Chicago, Pittsburgh) coalesced around industries that were based on mass production. The second phase is associated with the decline of the Fordist era and the rise of a post-Fordist flexible production. This form of industrial activity is based on small-size, small-batch units of (typically subcontracted) production that are nevertheless integrated into clusters of economic activity. Such clusters have been observed in two manifestations: labor-intensive craft forms (in Los Angeles, typically garments and jewelry) and high technology (especially the defense and aerospace industries). According to Scott, until recently, these so-called technopoles constituted the principal geographical loci of contemporary (sub)urbanization in Southern California.[9]

Post-Fordist regimes of accumulation are associated with analogous regimes of regulation or social control, most visible in the retreat from the welfare state. The rise of neoconservatism and the privatization ethos has coincided with a period of economic recession and retrenchment that has led many people to the brink of poverty just when the social welfare safety net is being withdrawn. In Los Angeles, as in many other cities, an acute socioeconomic polarization has resulted. In 1984, the city was dubbed the "homeless capital" of the United States because of the disproportionate concentration of homeless people (Wolch, 1990; Wolch & Dear, 1993; Wolch & Sommer, 1997).

For Soja (1996), restructuring in its broadest sense is central to Southern California's contemporary urban processes, giving rise to six distinct urban patterns. In addition to *exopolis* (noted earlier), Soja lists *flexcities,* associated with the transition to post-Fordism (especially deindustrialization and the rise of the information economy), and *cosmopolis,* referring to the globalization of Los Angeles in terms of both its emergent world city status and its internal multicultural diversification. According to Soja, peripheralization, post-Fordism, and globalization together define the experience of urban restructuring in Los Angeles. Three specific geographies result from these dynamics: *splintered labyrinth,* which describes the extreme forms of social, economic, and political polarization characteristic of the postmodern city; *carceral city,* referring to the new "incendiary urban geography" brought about by the amalgam of violence and police surveillance; and *simcities,* the term Soja uses to describe the new ways of seeing the city that are emerging from the study of Los Angeles—a type of epistemological restructuring that foregrounds a postmodern perspective.

Needless to say, any consideration of contemporary restructuring, whether of industrial production, population demographics, or urban landscapes, must encompass the globalization question sooner or later (cf. Knox & Taylor, 1995). In his reference to the global context of Los Angeles localisms, Davis (1992a) claims that if Los Angeles is in any sense paradigmatic, then it is because the city condenses the intended and unintended spatial consequences of post-Fordism. He insists that there is no simple master logic of restructuring, focusing instead on two key localized macroprocesses: the overaccumulation of bank and real estate capital in Southern California (principally from the East Asian trade surplus) and the reflux of low-wage manufacturing and labor-intensive service industries following on immigration from Mexico and Central America. For example, Davis notes how the City of Los Angeles used tax dollars gleaned from international capital investments to subsidize its downtown (Bunker Hill) urban renewal, a process he refers to as municipalized land speculation (p. 26). Through such connections, what happens today in Asia and Central America will have an effect in Los Angeles tomorrow. This global/local dialectic already has become an important (if somewhat imprecise) leitmotif of contemporary urban theory.

Ecodystopia

The natural environment of Southern California has been under constant assault since the first colonial settlements. Human habitation on a metropolitan scale has been possible only through a widespread manipulation of nature, especially the control of water resources in the American west (Davis, 1993; Gottlieb & FitzSimmons, 1991; Reisner, 1993). On the one hand, Southern Californians have a grudging respect for nature, living as they do adjacent to one of the earth's major geological hazards and in a desert environment that is prone to flood, landslide, and fire (see, e.g., Darlington, 1996; McPhee, 1989). On the other hand, its inhabitants have been energetically, ceaselessly, and sometimes carelessly unrolling the carpet of urbanization over the natural landscape for more than a century. This uninhibited occupation has engendered its own range of environmental problems, most notoriously air pollution but also issues related to habitat loss and dangerous encounters between humans and other animals.

The force of nature in Southern California has spawned a literature that attempts to incorporate environmental issues into the urban problematic. The politics of environmental regulation have long been studied in many places including Los Angeles (FitzSimmons & Gottlieb, 1996). However, the particular combination of circumstances in Southern California has stimulated an especially political view of nature, focusing both on its evisceration through human intervention (Davis, 1996) and on its potential for political mobilization by grassroots movements (Pulido, 1996). In addition, Wolch's (1996) Southern California–based research has led her to outline an alternative vision of biogeography's problematic.

Synthesis: Proto-Postmodern Urbanism

If these observers of the Southern California scene could talk with each other to resolve their differences and reconcile their terminologies,

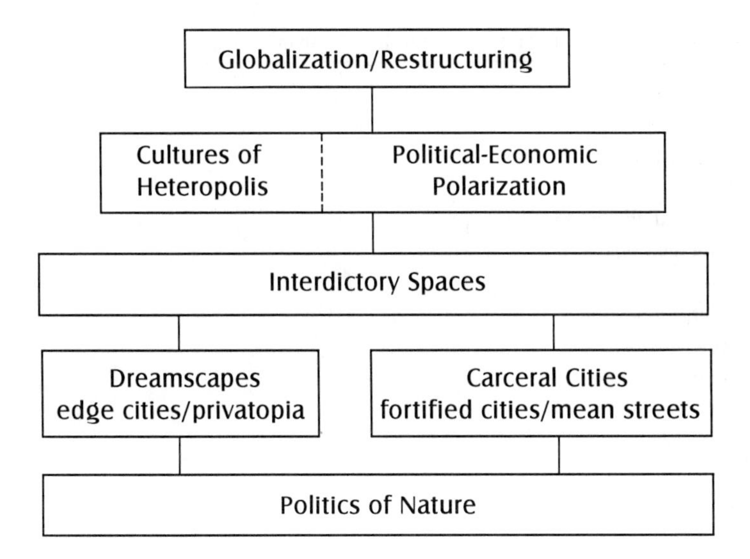

Figure 2.1. A Concept of Proto-Postmodern Urbanism

how might they do so? At the risk of misrepresenting their work, we suggest a schematic that is powerful yet inevitably incomplete (Figure 2.1). It suggests a proto-postmodern urban process driven by a global restructuring that is permeated and balkanized by a series of interdictory networks whose populations are socially and culturally heterogeneous but politically and economically polarized; whose residents are educated and persuaded to the consumption of dreamscapes even as the poorest are consigned to carceral cities; whose built environment, reflective of these processes, consists of edge cities, privatopias, and the like; and whose natural environment, also reflective of these processes, is being erased to the point of unlivability while at the same time providing a focus for political action.

■ Postmodern Urbanism

We anchor the postmodern urban problematic in the straightforward need to account for the evolution of society over time and space (Table 2.1). Such evolution occurs as a combination of deep-time (long-term) and present-time (short-term) processes, and it develops over

TABLE 2.1 Elements of a Postmodern Urbanism

Global latifundia
Holsteinization
Praedatorianism
Flexism
New world bipolar disorder
Cybergeoisie
Protosurps
Memetic contagion
Keno capitalism
Citistāt
Commudities
Cyburbia
Citadel
In-beyond
Cyberia
Pollyannarchy
Disinformation superhighway

several different scales of human activity (which we may represent summarily as micro-, meso-, and macroscales [Dear, 1988]). The structuring of the time-space fabric is the result of the interaction among ecologically situated human agents in relations of production, consumption, and coercion. We do not intend any primacy in this ordering of categories but instead emphasize their interdependencies; all are essential in explaining postmodern human geographies.

Our promiscuous use of neologisms in what follows is quite deliberate.[10] They are employed when no existing terms adequately describe the conditions we seek to identify, when a single term more conveniently substitutes for a complex phrase or string of ideas, and when neologistic novelty aids our avowed efforts to rehearse the break. The juxtaposing of postmodern and more traditional categories of modernist urbanism also is an essential piece of our analytical strategy. That modernist and postmodern categories overlap should surprise no one; we are, inevitably, building on existing urbanisms and epistemologies. The consequent neologistic pastiche may be properly regarded as a tactic of postmodern analysis or as analogous to hypothesis generation or the practice of dialectics.

Urban Pattern and Process

We begin with the assumption that urbanism is made possible by the exercise of instrumental control over both human and nonhuman ecologies. The very occupation and use of space, as well as the production and distribution of commodities, depends on an anthropocentric reconfiguration of natural processes and their products. As the scope and scale of, and dependency on, globally integrated consumption increases, institutional action converts complex ecologies into monocultured factors of production; nature is simplified into a *global latifundia.* This process includes both homogenizing interventions, as in California agriculture's reliance on vast expanses of single crops, and forceful interdiction to sustain that intervention against natural feedbacks, as in the aerial spraying of pesticides to eradicate fruit flies attracted to these vast expanses. Being part of nature, humanity is subjected to analogous dynamics.

Holsteinization is the process of monoculturing people as consumers to facilitate the harvesting of desires including the decomposition of communities into isolated family units and individuals to supplant social networks of mutual support with consumer sheds of dependent customers. Resistance is discouraged by means of *praedatorianism,* that is, the forceful interdiction by a praedatorian guard with varying degrees of legitimacy.

In one form or another, the global latifundia, holsteinization, and praedatorianism are as old as the global political economy, but the overarching dynamic signaling a break with previous manifestations is *flexism,* a pattern of econocultural production and consumption characterized by near instantaneous delivery and rapid redirectability of resource flows. Flexism's fluidity results from cheaper and faster systems of transportation and telecommunications, globalization of capital markets, and concomitant flexibly specialized, just-in-time production processes enabling short product cycles and production cycles. These result in highly mobile capital and commodity flows able to outmaneuver geographically fixed labor markets, communities, and bounded nation-states. Globalization and rapidity permit capital to evade long-term commitment to place-based socioeconomies. Whereas under Fordism exploitation is exercised through the alienation of labor in the place of production, flexism might require little or no labor at all from a given locale. Simultaneously, local down-waging and capital concentration operate synergistically to supplant locally owned enterprises with national and supranational chains, thereby transferring consumer capital and inventory selection ever further away from direct local control.

From these exchange asymmetries emerges a *new world bipolar disorder.* This globally bifurcated social order is many times more complicated than conventional class structures. In it, those overseeing the global latifundia enjoy concentrated power. Those who are dependent on command-and-control decisions find themselves in progressively weaker positions, pitted against each other globally and forced to accept shrinking compensation for their efforts (assuming that compensation is offered in the first place).

Of the two groups, the *cybergeoisie* reside in the "big house" of the global latifundia, providing indispensable, presently unautomatable command-and-control functions. They are predominantly stockholders, the core employees of thinned-down corporations, and write-your-own-ticket freelancers (e.g., chief executive officers, subcontract entrepreneurs, celebrities). They also may shelter members of marginal creative professions who comprise a type of *para-cybergeoisie.* The cybergeoisie enjoy perceived socioeconomic security and comparatively long-term horizons in decision making; consequently, their anxieties tend toward unforeseen social disruptions such as market fluctuations and crime. Commanding, controlling, and prodigiously enjoying the fruits of a shared global exchange of goods and information, the cybergeoisie exercise global coordination functions that predispose them to a similar ideology; thus, they are relatively heavily holsteinized.

Protosurps are the sharecroppers of the global latifundia. They are increasingly marginalized and surplus labor, providing just-in-time services when called on by flexist production processes but otherwise alienated from global systems of production (although not of consumption). Protosurps include temporary or day laborers, fire-at-will service workers, and a burgeoning class of intra- and international itinerant laborers specializing in pursuing the migrations of fluid investment. True *surpdom* is a state of superfluity beyond peonage—a vagrancy that is increasingly criminalized through anti-homeless ordinances, welfare state erosion, and widespread community intolerance (of, e.g., all forms of panhandling). Protosurps are called on to provide as yet unautomated service functions that can be performed by anyone. Subjected to high degrees of uncertainty by the omnipresent threat of instant unemployment, protosurps are prone to clustering into affinity groups for support in the face of adversity. These affinity groups, however, are not exclusive, overlapping in both membership and space, resulting in a class of marginalized indigenous populations and peripheral immigrants who are relatively less holsteinized.

The sociocultural collisions and intermeshings of protosurp affinity groups, generated by flexist-induced immigration and severe social differentiation, produce wild *memetic contagion.*[11] This is a process by which cultural elements of one individual or group exert crossover influences on the culture of another previously unexposed individual or group. Memetic contagion is evidenced in Los Angeles by hybridized agents and intercultural conflicts such as Mexican and Central American practitioners of Afro-Caribbean religion (McGuire & Scrymgeour, 1998), blue-bandanna'd Thai Crips, and the adjustments prompted by poor African Americans' offense at Korean merchants' disinclination to smile casually. Memetic contagion should not be taken for a mere epiphenomenon of an underlying political economic order, generating colorfully chaotic ornamentation for a flexist regime. Rather, it entails the assemblage of novel ways of seeing and being from whence new identities, cultures, and political alignments emerge. In turn, these new social configurations may force change in existing institutions and structures and may spawn cognitive conceptions that are incommensurable with, although not necessarily any less valid than, existing models. The inevitable tensions between the anarchic diversification born of memetic contagion and the manipulations of the holsteinization process might yet prove to be the central cultural contradiction of flexism.

With the flexist imposition of global imperatives on local economies and cultures, the spatial logic of Fordism has given way to a new, more dissonant international geographical order. In the absence of conventional communication and transportation imperatives mandating propinquity, the once standard Chicago School logic has given way to a seemingly haphazard juxtaposition of land uses scattered over the landscape. Worldwide, agricultural lands sprout monocultures of exportable strawberries or broccoli in lieu of diverse staple crops grown for local consumption. Sitting amid these fields, identical assembly lines produce the same brand of automobile, supplied with parts and managed from distant continents. Expensive condominiums appear among squatter slums, indistinguishable in form and occupancy from (and often in direct communication with) luxury housing built atop homeless encampments elsewhere in the world.

Yet, what close up appears to be a fragmented, collaged polyculture is, from a longer perspective, a geographically disjoint but hyperspatially integrated monoculture, that is, shuffled similarities set amid adaptive and persistent local variations. The result is a landscape not unlike that formed by a keno game card. The card itself appears as a numbered grid, with some squares being marked during the course of the game and others not marked, according to a random draw. The process governing this marking, determined by a rationalized set of procedures beyond the territory of the card itself, ultimately determines which player will achieve a jackpot-winning pattern. Similarly, the apparently random development and redevelopment of urban land may be regarded as the outcome of exogenous investment processes inherent to flexism, thus creating the landscapes of *keno capitalism.*

Keno capitalism's contingent mosaic of variegated monocultures renders discussion of "the city" increasingly reductionist. More holistically, the dispersed net of megalopoles may be viewed as a single integrated urban system or *Citistāt.* Citistāt, the collective world city, has emerged from competing urban webs of colonial and postcolonial eras to become a geographically diffuse hub of an omnipresent periphery, drawing labor and materials from readily substitutable locations throughout that periphery. It is both geographically corporeal, in the sense that urban places exist, and yet geographically ethereal, in the sense that communication systems create a virtual space permitting coordination across physical space. Both realms reinforce each other while (re)producing the new world bipolar disorder.

Materially, Citistāt consists of *commudities* (commodified cybergeois residential and commercial ecologies) and the *in-beyond* (internal peripheries simultaneously undergoing but resisting instrumentalization in myriad ways). Virtually, Citistāt consists of *cyburbia,* the collection of state-of-the-art data transmission, premium pay-per-use, and interactive services generally reliant on costly and technologically complex interfaces, and *cyberia,* an electronic outland of rudimentary communications including basic phone service and telegraphy, interwoven with and preceptorally conditioned by the disinformation superhighway (DSH).

Commudities are commodified communities created expressly to satisfy (and profit from) the habitat preferences of the well-recompensed cybergeoisie. They commonly consist of carefully manicured residential and commercial ecologies managed through privatopian self-administration and maintained against internal and external outlaws by a repertoire of interdictory prohibitions. Increasingly, these prepackaged environments jockey with one another for clientele on the basis of recreational, cultural, security, and educational amenities. Commonly located on difficult-to-access sites such as hilltops and urban edges, far from restless populations undergoing conversion to protosurpdom, individual commudities are increasingly teleintegrated to form cyburbia (Dewey, 1994), the interactive tollways comprising the high-rent district of Citistāt's hyperspatial electronic shadow.[12] Teleintegration already is complete (and de rigueur) for the *citadels,* which are commercial commudities consisting of high-rise corporate towers from which the control and coordination of production and distribution in the global latifundia is exercised.

Citistāt's internal periphery and repository of cheap on-call labor lies at the *in-beyond,* composed of a shifting matrix of protosurp affinity clusters. The in-beyond may be envisioned as a patchwork quilt of variously defined interest groups (with differing levels of economic, cultural, and street influence), none of which possesses the wherewithal to achieve hegemonic status or to secede. Secession may occur locally to some degree, as in the cases of the publicly subsidized reconfiguration of Los Angeles's Little Tokyo and the consolidation of Koreatown through the import, adjacent extraction, and community recirculation of capital. The piecemeal diversity of the in-beyond makes it a hotbed of wild memetic contagion.

The global connectivity of the in-beyond is considerably less glamorous than that of the cybergeoisie's commudities, but it is no less extensive. Intermittent phone contact and wire service remittances occur throughout cyburbia (Rushkoff, 1995; also see Knox & Taylor, 1995). The potholed public streets of Citistāt's virtual twin are augmented by extensive networks of snail mail, personal migration, and the hand-to-hand passage of mediated communications (e.g., cassette tapes). Such contacts occasionally diffuse into commudities.

Political relations in Citistāt tend toward polyanarchy, a politics of grudging tolerance of "difference" that emerges from interactions and accommodations within the in-beyond and between commudities and, less frequently, between in-beyond and commudity. Its more pervasive form is *pollyannarchy,* an exaggerated, manufactured optimism that promotes a self-congratulatory awareness and respect for difference and the asymmetries of power. Thus, pollyannarchy is a pathological form of polyanarchy, disempowering those who would challenge the controlling beneficiaries of the new world bipolar disorder. Pollyannarchy is evident in the continuing spectacle of electoral politics and in the citywide unity campaign run by corporate sponsors following the 1992 uprising in Los Angeles.

Wired throughout the body of Citistāt is the DSH, a mass infotainmercial media owned by roughly two dozen cybergeoisie institutions. The DSH disseminates holsteinizing ideologies and incentives, creates wants and dreams, and inflates the symbolic value of commodities. At the same time, it serves as the highly filtered sensory organ through which commudities and the in-beyond perceive the world outside their unmediated daily experiences. The DSH is Citistāt's "consent factory" (Chomsky & Herman, 1988), engineering memetic contagion to encourage participation in a global latifundia that is represented as both inevitable and desirable. Because the DSH is a broadband distributor of information designed primarily to attract and deliver consumers to advertisers, the ultimate reception of messages is difficult to target and predetermine.

Thus, the DSH serves inadvertently as a vector for memetic contagion such as the conversion of cybergeoisie youth to wannabe gangstas via the dissemination of hip-hop culture over commudity boundaries. The DSH serves as a network of preceptoral control and, therefore, is distinct from the coercive mechanisms of the praedatorian guard. Overlap between the two is increasingly common, however, as in the case of televised disinfotainment programs such as *America's Most Wanted,* in which crimes are dramatically reenacted and viewers are invited to call in and betray alleged perpetrators.

As the cybergeoisie increasingly withdraw from the Fordist redistributive triad of big government, big business, and big labor to establish their own micro-nations, the social support functions of the state disintegrate, along with the survivability of less affluent citizens. The global migrations of work to the lowest wage locations of the in-beyond and of consumer capital to the citadels result in power asymmetries that become so pronounced that even the DSH is at times incapable of obscuring them, leaving protosurps increasingly disinclined to adhere to the remnants of a tattered social contract. This instability creates the potential for violence, pitting Citistāt and cybergeoisie against the protosurp in-beyond and inevitably leading to a demand for the suppression of protosurp intractability. Thus, the praedatorian guard emerges as the principal remaining vestige of the police powers of the state. This increasingly privatized public/private partnership of mercenary sentries, police expeditionary forces, and their technological extensions (e.g., video cameras, helicopters, criminological data uplinks) watches over the commudities and minimizes disruptiveness by acting as a force of occupation within the in-beyond. The praedatorian guard achieves control through coercion, even at the international level, where asymmetrical trade relations are reinforced by the military and its clientele. It might be only a matter of time before the local and national praedatorians are administratively and functionally merged, as exemplified by proposals to deploy military units for policing inner-city streets or the U.S.-Mexico border.

■ Conclusion: Invitation to a Postmodern Urbanism

In interrogating the ever expanding body of literature addressing the urban landscape of Los Angeles, we have pieced together an urban model grounded in the lived experiences and built forms of the end of the 20th century and, perhaps, the beginning of the next century. We do not pretend to have completed this project, nor do we claim that the Southern California experience is necessarily typical of other metropolitan regions in the United States or the world. Still less would we advocate that our model, or any other model, be established as a newly hegemonic vision of the urban; to do so would be to violate the rejection of master narratives critical to a postmodern ontology. In the absence of existing urban models congruent with contemporary urban conditions, we have assembled a model that, although necessarily fragmented, partial, positional, and awaiting dialogical engagement with alternate conceptions of the urban from within and beyond Los Angeles, recognizes and describes a postmodern urban process in which the urban periphery organizes the center in the context of a globalizing capitalism.

This postmodern urban process remains resolutely capitalist, even though the nature of that enterprise is changing in very significant ways, especially through, for example, the telecommunications revolution, the changing nature of work, and globalization. A radical break is occurring, this time in the conditions of our material world. Contemporary urbanism is a consequence of how local and interlocal flows of material and information (including symbols) intersect in a rapidly converging, globally integrated economy driven by the imperatives of flexism. Landscapes and peoples are homogenized to facilitate large-scale production and consumption. Highly mobile capital and commodity flows outmaneuver geographically fixed labor markets, communities, and nation-states and cause a globally bifurcated polarization. The beneficiaries of this system are the cybergeoisie, even as the numbers of permanently marginalized protosurps grow.

In the new global order, socioeconomic polarization and massive, sudden migrations spawn cultural hybrids through the process of memetic contagion. Cities no longer develop as concentrated loci of population and economic activity; instead, they develop as fragmented parcels within Citistāt, the collective world city. Materially, Citistāt consists of commudities (commodified communities) and the in-beyond (the permanently marginalized). Virtually, Citistāt is composed of cyburbia (those who are hooked into the electronic world) and cyberia (those who are not). Social order is maintained by the ideological apparatus of the DSH, Citistāt's consent factory, and by the praedatorian guard, the privatized vestiges of the nation-state's police powers.

Keno capitalism is the synoptic term that describes the spatial manifestations of the postmodern urban condition. Urbanization is occurring on a quasi-random field of opportunities. Capital touches down as if by chance on a parcel of land, ignoring the intervening opportunities and sparking the development process. The relationship between development of one parcel and nondevelopment of another is a disjointed, seemingly unrelated affair. Although not truly a random process, the traditional, center-driven agglomeration economies that have guided urban development in the past no longer apply. Conventional city form, Chicago style, is sacrificed in favor of a noncontiguous collage of parcelized, consumption-oriented landscapes devoid of conventional centers yet wired into electronic propinquity and nominally unified by the mythologies of the DSH.

Los Angeles might be a mature form of this postmodern metropolis; Las Vegas comes to mind as a youthful example. The consequent urban aggregate is characterized by acute fragmentation and specialization—a partitioned gaming board subject to perverse laws and peculiarly discrete, disjointed urban outcomes. Given the pervasive presence of crime, corruption, and violence in the global city (not to mention geopolitical transitions as nation-states give way to micro-nationalisms and transnational mafias), the city as gaming board seems an especially appropriate 21st-century successor to the concentrically ringed city of the early 20th century.

We intend this alternative model of urban structure as an invitation to examine the concept of a postmodern urbanism. We have only begun to sketch its potential; its validity will be properly assessed only if researchers elsewhere in the world are willing to examine its precepts. We urge others to share in this enterprise. Even though our vision is tentative, we are convinced that we have glimpsed a new way of understanding cities.[13]

NOTES

1. See, for example, Auge (1995) and Pred (1995).

2. Some elements of this discussion may be found in Ellin (1996), Knox and Taylor (1995), and Watson and Gibson (1996).

3. The theoretical bases for this argument are examined more fully in Dear (1988, 1991). For specific considerations of the rhetoric of city planning in the new urbanism, see Dear (1989).

4. Raban's view is echoed in the seminal work of de Certeau (1984).

5. Other studies of suburbanization in Los Angeles, most notably by Hise (1997) and Waldie (1996), provide a basis for comparing planned community marketing in Southern California.

6. Such sentiments are echoed in Smith's (1992) assessment of the new urban frontier, where expression is powered by two industries: real estate developers (who package and define value) and the manufacturers of culture (who define taste and consumption preferences) (p. 75).

7. The list of Los Angeles novels and movies is endless. Typical of the dystopian cinematic vision are *Blade Runner* (1986) and *Chinatown* (1974) and of silly optimism is *L.A. Story* (1991).

8. One very important feature of local neighborhood dynamics in the fortified culture of Southern Californian cities is the presence of street gangs (Klein, 1995; Vigil, 1988).

9. This development was prefigured in Fishman's (1987) description of the "tech-noburb." See also Castells and Hall (1994).

10. One critic accused us, quite cleverly, of "neologorrhea."

11. This term is a combination of Girard's "mimetic contagion" and Dawkins's (1996) hypothesis that cultural information consists of gene-type units, or "memes," transmitted virus-like from head to head. We employ the term *hybridized* in recognition of the recency and novelty of the combination, not to assert prior purity to the component elements forming the hybrid.

12. This proccess might soon find a geographical analog in the conversion of automotive freeways into exclusive tollways linking commudities.

13. The collection of essays assembled in Benko and Strohmayer (1997) is an excellent overview of the relationship between space and postmodernism, including the urban question. Robins' valuable work on media, visual cultures, and representational issues also deserves a wide audience (Morley & Robins, 1995; Robins, 1996).

REFERENCES

Abelmann, N., & Lie, J. (1995). *Blue dreams: Korean Americans and the Los Angeles riots.* Cambridge, MA: Harvard University Press.

Anderson, S. (1996). A city called heaven: Black enchantment and despair in Los Angeles. In A. J. Scott & E. Soja (Eds.), *The city: Los Angeles and urban theory at the end of the twentieth century* (pp. 336-364). Los Angeles: University of California Press.

Auge, M. (1995). *Non-places: Introduction to an anthropology of super-modernity.* London: Verso.

Baldassare, M. (Ed.). (1994). *The Los Angeles riots.* Boulder, CO: Westview.

Banham, R. (1973). *Los Angeles: The architecture of four ecologies.* London: Penguin .

Benko, C., & Strohmayer, U. (Eds.). (1997). *Space and social theory: Interpreting modernity and postmodernity.* Oxford, UK: Blackwell.

Boyd, T. (1996). A small introduction to the "G" funk era: Gangsta rap and black masculinity in contemporary Los Angeles. In M. Dear, H. E. Schockman, & G. Hise (Eds.), *Rethinking Los Angeles* (pp. 127-146). Thousand Oaks, CA: Sage.

Boyd, T. (1997). *Am I black enough for you?* Bloomington: University of Indiana Press.

Bullard, R. D., Grigsby, J. E., & Lee, C. (1994). *Residential apartheid.* Los Angeles: UCLA Center for Afro-American Studies.

Butler, O. E. (1993). *Parable of the sower.* New York: Four Walls Eight Windows.

Castells, M., & Hall, P. (1994). *Technopoles of the world.* New York: Routledge.

Chomsky, N., & Herman, E. (1988). *Manufacturing consent.* New York: Pantheon .

Darlington, D. (1996). *The Mojave: Portrait of the definitive American desert.* New York: Henry Holt.

Davis, M. (1992a). Chinatown revisited? The internationalization of downtown Los Angeles. In D. Reid (Ed.), *Sex, death, and God in L.A.* (pp. 54-71). New York: Pantheon.

Davis, M. (1992b). Fortress Los Angeles: The militarization of urban space. In M. Sorkin (Ed.), *Variations on a theme park* (pp. 154-180). New York: Noonday.

Davis, M. (1996). How Eden lost its garden: A political history of the Los Angeles landscape. In A. J. Scott & E. Soja (Eds.), *The city: Los Angeles and urban theory at the end of the twentieth century* (pp. 160-185). Los Angeles: University of California Press.

Davis, M. L. (1993). *Rivers in the desert: William Mulholland and the invention of Los Angeles.* New York: HarperCollins.

Dawkins, R. (1996). *The selfish gene.* New York: Oxford University Press.

Dear, M. (1988). The postmodern challenge: Reconstructing human geography. *Transactions: Institute of British Geographers, 13,* 262-274.

Dear, M. (1989). Privatization and the rhetoric of planning practice. *Society & Space, 7,* 449-462.

Dear, M. (1991). The premature demise of postmodern urbanism. *Cultural Anthropology, 6,* 538-552.

de Certeau, M. (1984). *The practice of everyday life.* Berkeley: University of California Press.

Dewey, F. (1994, May). Cyburbia: Los Angeles as the new frontier, or grave? *Los Angeles Forum for Architecture and Urban Design Newsletter,* pp. 6-7.

Ellin, N. (1996). *Postmodern urbanism.* Oxford, UK: Blackwell.

Fishman, R. (1987). *Bourgeois utopias: The rise and fall of suburbia.* New York: Basic Books.
FitzSimmons, M., & Gottlieb, R. (1996). Bounding and binding metropolitan space: The ambiguous politics of nature in Los Angeles. In A. J. Scott & E. Soja (Eds.), *The city: Los Angeles and urban theory at the end of the twentieth century* (pp. 186-224). Los Angeles: University of California Press.
Flusty, S. (1994). *Building paranoia: The proliferation of interdictory space and the erosion of spatial justice.* West Hollywood, CA: Los Angeles Forum for Architecture and Urban Design.
Garreau, J. (1991). *Edge city: Life on the new frontier.* New York: Doubleday.
Gibson, W. (1996). *Idoru.* New York: Putnam.
Gooding-Williams, R. (Ed.). (1993). *Reading Rodney King, reading urban uprising.* New York: Routledge.
Gottlieb, R., & FitzSimmons, M. (1991). *Thirst for growth: Water agencies and hidden government in California.* Tucson: University of Arizona Press.
Hise, G. (1997). *Magnetic Los Angeles: Planning the twentieth-century metropolis.* Baltimore, MD: Johns Hopkins University Press.
Jencks, C. (1993). *Heteropolis: Los Angeles, the riots, and the strange beauty of hetero-architecture.* London: Academy Editions.
Klein, M. (1995). *The American street gang.* New York: Oxford University Press.
Knox, P., & Taylor, P. J. (Eds.). (1995). *World cities in a world system.* Cambridge, UK: Cambridge University Press.
Martinez, R. (1992). *The other side: Notes from the new L.A., Mexico City, and beyond.* New York: Vintage.
McGuire, B., & Scrymgeour, D. (1998). Santeria and Curanderismo in Los Angeles. In P. Clarke (Ed.), *New trends and developments in African religion* (pp. 211-222). Wesport, CT: Greenwood.
McKenzie, E. (1994).*Privatopia: Homeowner associations and the rise of residential private government.* New Haven, CT: Yale University Press.
McPhee, J. (1989). *The control of nature.* New York: Noonday.
Mills, C. W. (1959). *The sociological imagination.* New York: Oxford University Press.
Molotch, H. (1996). L.A. as design product: How art works in a regional economy. In A. J. Scott & E. Soja (Eds.), *The city: Los Angeles and urban theory at the end of the twentieth century* (pp. 225-275). Los Angeles: University of California Press.
Morley, D., & Robins, K. (1995). *Spaces of identity: Global media, electronic landscapes, and cultural boundaries.* New York: Routledge.
Ong, P., Bonacich, E., & Cheng, L. (Eds.). (1994). *The new Asian immigration in Los Angeles and global restructuring.* Philadelphia: Temple University Press.
Park, E. (1996). Our L.A.? Korean Americans in Los Angeles after the civil unrest. In M. Dear, H. E. Schockman, & G. Hise (Eds.), *Rethinking Los Angeles* (pp. 153-168). Thousand Oaks, CA: Sage.
Pred, A. (1995). *Recognizing European modernities.* New York: Routledge.
Pulido, L. (1996). Multiracial organizing among environmental justice activists in Los Angeles. In M. Dear, H. E. Schockman, & G. Hise (Eds.), *Rethinking Los Angeles* (pp. 171-189). Thousand Oaks, CA: Sage.
Raban, J. (1974). *Soft city.* New York: E. P. Dutton.
Reisner, M. (1993). *Cadillac desert: The American west and its disappearing water.* New York: Penguin.

Relph, E. C. (1987). *The modern urban landscape.* Baltimore, MD: Johns Hopkins University Press.

Robins, K. (1996). *Into the image: Culture and politics in the field of vision.* New York: Routledge.

Rocco, R. (1996). Latino Los Angeles: Reframing boundaries/borders. In A. J. Scott & E. Soja (Eds.), *The city: Los Angeles and urban theory at the end of the twentieth century* (pp. 365-389). Los Angeles: University of California Press.

Roseman, C., Laux, H. D., & Thieme, G. (Eds.). (1996). *Ethnicity.* Lanham, MD: Rowman & Littlefield.

Rushkoff, D. (1995). *Cyberia: Life in the trenches of hyperspace.* New York: HarperCollins.

Scott, A. J. (1988a). *Metropolis: From the divison of labor to urban form.* Berkeley: University of California Press.

Scott, A. J. (1988b). *New industrial spaces: Flexible production, organization and regional development in North America and Western Europe.* London: Pion.

Scott, A. J. (1993). *Technopolis: High-technology industry and regional development in Southern California.* Berkeley: University of California Press.

Smith, N. (1992). New city, new frontier. In M. Sorkin (Ed.), *Variations on a theme park* (pp. 61-93). New York: Noonday.

Soja, E. (1989). *Postmodern geographies: The reassertion of space in critical social theory.* London: Verso.

Soja, E. (1992). Inside exopolis: Scenes from Orange County. In M. Sorkin (Ed.), *Variations on a theme park* (pp. 94-122). New York: Noonday.

Soja, E. (1996). Los Angeles 1965-1992: The six geographies of urban restructuring. In A. J. Scott & E. Soja (Eds.), *The city: Los Angeles and urban theory at the end of the twentieth century* (pp. 426-462). Los Angeles: University of California Press.

Sorkin, M. (Ed.). (1992). *Variations on a theme park.* New York: Noonday.

Storper, M., & Walker, R. (1989). *The capitalist imperative.* Oxford, UK: Blackwell.

Suisman, D. R. (1989). *Los Angeles Boulevard.* West Hollywood, CA: Los Angeles Forum for Architectural and Urban Design.

Vigil, J. (1988). *Barrio gangs: Streetlife and identity in Southern California.* Austin: University of Texas Press.

Waldie, D. J. (1996). *Holy land: A suburban memoir.* New York: Norton.

Waldinger, R., & Bozorgmehr, M. (1996). *Ethnic Los Angeles.* New York: Russell Sage.

Watson, S., & Gibson, K. (Eds.). (1996). *Postmodern cities and spaces.* Oxford, UK: Blackwell.

Wolch, J. (1990). *The shadow state: Government and voluntary sector transition.* New York: Foundation Center.

Wolch, J. (1996). From global to local: The rise of homelessness in Los Angeles during the 1980s. In A. J. Scott & E. Soja (Eds.), *The city: Los Angeles and urban theory at the end of the twentieth century* (pp. 390-425). Los Angeles: University of California Press.

Wolch, J., & Dear, M. (1993). *Malign neglect: Homelessness in an American city.* San Francisco: Jossey-Bass.

Wolch, J., & Sommer, H. (1997). *Los Angeles in an era of welfare reform.* Los Angeles: Liberty Hill Foundation.

Yoon, I. (1997). *On my own: Korean businesses and race relations in America.* Chicago: University of Chicago Press.

3

Crossing CyberCities: Urban Regions and the Cyberspace Matrix

M. CHRISTINE BOYER

The dominant metaphor of cyberspace is the matrix, the web, the net, the lattice, and the field. It includes at its core the virtual matrices—the bifurcating and diverging of lines and links between points in time and space that define the Internet (or Net). On the periphery is either the space between virtual and actual worlds or those leftover interstitial spaces between the nodes of the matrix, that is, spaces pushed through or left out of the Net.

Elaborating on the analogy between the virtual space of the computer matrix and the material space of physical cities, described in my *CyberCities* (Boyer, 1996), this chapter moves back and forth between virtual and actual reality to explore the concatenated term of *cybercities*. *CyberCities* probes the relationship between the imaginary real space of users of computer-mediated information and the spatial and temporal experiences of city users. It asks how technological devices such as the computer and its representational metaphors such as the matrix alter perception and direct the formation of knowledge.

The analogy between the computer matrix and the city relies on how users organize space, lay down routes by which to navigate this space, and build cognitive maps. There are differences, however. Cyberspace has no way of treating boundaries, no way of enabling users to cross over the threshold separating virtual from actual reality. This problem exists with any mathematicization of space and time, with any modeling of real space and time that involves the formal applicability of conceptual tools (Markley, 1996).

The Internet is used in this chapter as the prime example of the electronic matrix of cyberspace. The chapter argues that although the relationship between this electronic matrix and the regional city is only an analogy, it is a strong one that needs to be explored for its hidden assumptions. More specifically, the chapter examines the question of democracy on the Net and the new development of virtual public space by pointing to a digital divide between those connected electronically and those in lag-time places. The chapter also focuses on the nodes of the matrix, frame-like spaces cut from the rest of the city and operating as though they were computer algorithms. With their design codes and simulation rules, these decorated spaces reinforce the perception that the regional space of the city is a matrix of well-defined nodes, with interstitial jump-cut spaces left out of the perceptual frame. To draw attention to the distinction between symbolic and associative processes as they map out the territory of cybercities, I also examine the issue of cognitive mapping and memory devices. Finally, I cast the space of the Net as a linguistic terrain and note how electronic writing has begun to move across architectural and urban space.

From the moment that William Gibson announced in his dystopic science fiction *Neuromancer* that "cyberspace" looks like Los Angeles seen from 5,000 feet in the air (Gibson, 1984), there has been a predilection for drawing parallels between the virtual space of computer networks and posturban places of disorder and decay. The cities in our minds are described as a series of lights receding into the distance, a myriad of connections spreading horizontally and invisibly. Bruce Sterling's *Islands in the Net* reworks Gibson's concept of the matrix:

> Every year in her life, Laura thought, the Net had been growing more expansive and seamless. Computers did it. Computers melted other machines, fusing them together. Television-telephone-telex. Tape recorder-VCR-laser disk. Broadcast tower linked to microwave dish linked to satellite. Phone line, cable TV, fiber-optic cords hissing out words and pictures in torrents of pure light. All netted together in a web over the world, a global nervous system, an octopus of data. (Sterling, 1989, p. 17; also quoted in Fitting, 1991, p. 229)

The analogy also has been reversed, with cyberspace being called a huge megalopolis without a center, both a city of regional sprawl and an urban jungle. Indeed, what else is the American city of today but a gigantic, boundless metroscape like "BAMA" (Boston to Atlanta)? Its appearance seems to simulate a complex switchboard of plug-in zones and edge cities connected through an elaborate network of highways, telephones, computer banks, fiber-optic cable lines, and television and radio outlets. The metaphoric translation of superhighways into the "information highway" is a prime example of how this analogy of cybercities determines perception and hides a set of assumptions. Gibson's Ur-matrix "I-95" is in reality the product of massive investments of capital and labor in federally funded highway-building programs plus the post-World War II boom in automobile production. Enormous costs were produced in their wake—fossil fuel pollution, neglect of public transit, environmental degradation, and community destruction. Yet, the metaphor of the information highway glibly erases this complex set of costs, assuming a technological mind-set that we can remake the world without paying attention to consequences.

Although the task of bringing America online will not generate the same set of charges, it will determine a new set of power-knowledge relationships that should be taken into account. Presented as a natural development of science and technology, the narrative logic of the information highway assumes that we can extrapolate from the present into the future based on the evolutionary progress in software and hardware technologies that inevitably will lead to revolutionary improvement (Markley, 1996).

Drawing an analogy between cyberspace and the regional city settles Christopher Alexander's argument that "A City Is Not a Tree." Alexander (1965) felt that a tree structure, representing a hierarchical and linear arrangement of divisions (or bifurcations) dictating the organization of knowledge or urban places, was too rigid and reductive. He proposed that the space of the city be reconceptualized as a semi-lattice. Lacking hierarchy, each point in this abstract structure could be linked to any other point, not just to the branches above or below it. Instantaneous accessibility and connectivity replace the hierarchical pathways of tree structures. Gilles Deleuze and Felix Guattari concur. They criticize the tree structures that have dominated Western thought from botany to philosophy: "We're tired of trees. We should stop believing in trees, roots, and radicles. They've made us suffer too much. All of arborescent culture is founded on them, from biology to linguistics" (Deleuze & Guattari, 1987, quoted in Ostwald, 1997, p. 452).

The concept of the semi-lattice does away with center-periphery arguments and the notion of sub-urbs that depend on a central urb. A lattice or net is open-ended, potentially connected to any point on its grid. As it spreads laterally throughout an invisible ethereal space, it is capable of generating infinite complexity.

The Democratic Peformance of Cyberspace

Thinking of cybercities as a net covers up a multitude of erasures and avoidances. What does it mean that this electronic imagery of the matrix generates a unique mental ordering that parallels, but does not represent, reality? What significant effects result from the fact that textual universes of postmodern accounts conjure up fictional worlds that disavow any link with material reality, any connectivity with a shared community? Of course, "shared community" is open to criticism; perhaps it is too nostalgic for a lost public sphere or assumes erroneously that democracy is based on Enlightenment principles of open discussion, normative behavior, and rational purposive action.

We need to ask whether old concepts of freedom of speech and individual rights ruled by law are sufficient for new forms of electronic communication, public assemblage, and the decentralized dialogues taking place across the globe on the Net. The Internet is a decentered communications system, a product of cold war policies that included nonhierarchical communications networks that would survive a nuclear attack if their hub was eliminated (Guy, n.d.). Now, the Net consists of a network of networks, a heterotopia of discourses that exists as pixels on a screen generated at remote locations by individuals who probably never will meet face-to-face. What does this say about a *shared community*? Does the Internet constitute an increase in democracy, as many suggest?

Anyone can access the Net from a computer terminal if he or she pays a commercial provider for the service, but central authorities still have to recognize that name space for others to find it. This entry point to the Net is controlled in America by two private companies: AT&T, the telecommunications giant, and Scientific Applications International Corporation (SAIC), a firm that specializes in military, police, and government intelligence contracts.[1] Albeit a nonhierarchical cooperative mode of communication, the Net is highly centralized and controlled. For example, the federal government commands the root server A, which distributes any new address information to other root servers around the world. Furthermore, a private company, Network Solutions

Inc., was awarded an exclusive government contract in 1995 to administer this crucial root server and to register domain names such as `.com`, `.org`, and `.net`. This profitable monopoly has enabled Network Solutions to generate millions of dollars in profits from registration fees as the Internet itself has grown bigger and bigger. Such monopoly control flies in the face of those who believe the Internet to be an open-ended, cooperative, and egalitarian enterprise (Harmon, 1998).

Although the Net often is referred to as an information highway, will it be a toll road or freeway? Will the traffic that streams across it be controlled or not? Will the market dominate the Net? Will it be commercialized or free for those who can afford a telephone and computer? How do the answers to these questions relate to physical space?

As formerly independent telecommunications, computer, media, and entertainment industries merge into transnational communication corporations, they have set their sights on controlling the global information and entertainment network. The chief executive officer of Walt Disney Company has suggested that his company's prospective growth is based on the fact that its "nonpolitical product" does not threaten any political regime around the world (McChesney, 1996, p. 115). Meanwhile, public sector communications are being dismantled, and their independent voices—both journalistic and cultural—are increasingly quiet. A free flow of information is essential for a lively democratic political culture, but will uncontrolled networks of information—with universal access and open discussion—survive without some form of government or market regulation or without some form of public scrutiny? What should that form of regulation or scrutiny be?

In spite of what many advocates of cyberspace believe, universal access is far from guaranteed. Estimates made in 1996 claim that more than 25 million people communicate on the Internet, a figure that grows yearly. Yet, a 1995 estimate asserted that although one third of U.S. households owned computers, many of them had no access to the Internet and, in poor neighborhoods, at least one third of the households were without basic telephone connections (Morris & Ogan, 1996). Hence, the computer could become another device that increases economic and educational disparity.

According to Mitchell L. Moss and Anthony Townsend of New York University's Taub Urban Research Center, 50% of all Internet hosts that connect computers to the Internet are located in just five states and are concentrated in a few metropolitan regions within these five states.

Furthermore, 30% of these hosts are clustered along the Northeast Corridor between Maine and Virginia (not quite BAMA). If California is added, then nearly half of all the hosts have been included. Internet hosts also are concentrated in certain regional cities. Uppermost is Silicon Valley outside of San Francisco, then Route 128 (the high-tech corridor outside of Boston). Next comes Los Angeles County. Manhattan continues to dominate its region due to the enormous number of information-intensive financial and media businesses centered there. A number of large cities have no Internet hosts—Houston, Miami, Detroit, and New Orleans, for example. These are information black holes, and no doubt these cities' economic growth and well-being will be inhibited in the 21st century's information society (Moss & Townsend, 1996).

Stephen Doheny-Farina, in *The Wired Neighborhood,* explores the creation of electronic neighborhoods and the virtualization of community. He is critical of the absolute dependence that these new communities place on a technology that is created, provided, and sustained by others. Rather than the euphoria of technological liberation, independence, and self-sufficiency, he sees these virtual communities as a sign of containment and domestication (Doheny-Farina, 1996).[2] We do not know for certain the real effects—whether virtual communities strengthen social ties, provide community education, or counteract the fragmenting tendency of privatization. Can they reduce the cost of delivering government services or update education and bring children into the 21st century regardless of the overall quality of the school system, income levels of households, geographic location, and ethnic backgrounds?

Other critics complain that the new telecommunications technology centralizes control over cultural production and homogenizes cultural consumption by peddling the same cultural products worldwide. Moreover, it makes government irrelevant. They argue that community networks will lead to an information dystopia—isolated and alienated citizens locked together within their electronic prisons while a powerful technical elite gains actual mind control through panopticon surveillance. Silent surveillance results as communications are scanned for selected keywords and then stored as a virtual profile characterizing each discussant.

A more optimistic approach to the new democratic and egalitarian Net is Neil Guy's *Community Network* (Guy, 1991). He studied computer-based communication systems called *free-nets* or *public access networks* created out of community computing centers, civic networks, and telecommunities. Owned by nonprofit organizations, these networks use the Internet for community welfare. They provide an open forum for citizen discussions and bring fractured communities together. This connectivity enables communities to perceive things in different ways; people know more about what is going on, and they change their behavior in relationship to each other. Most community organizers, however, rely on face-to-face contacts, and some are not computer literate (Guy, n.d.), thereby setting limits on these virtual communities.

A mounting digital divide also exists between those connected to and those disconnected from the electronic matrix. Because they have yet to make their first telephone calls, two thirds of the world's population is excluded from cyberspace. Even if one is connected electronically, the Net is predominantly an English-speaking realm, and this also limits accessibility. Just look at the facts and figures; fully 90% of all Net users live in North America and Europe because 80% of the world lacks basic telecommunication devices (i.e., telephones). Access to the Net is 12 times more expensive in Indonesia than in Rome.

Investors are unwilling to share the risk of developing telephone service in Africa, for example, because of political instability in much of the continent (Landler, 1995). Of course, the hope is held out that the Internet would offer "a great leap forward in Africa," the poorest continent, but Africa is crippled by unreliable communication systems, on the one hand, and prohibitively expensive telecommunications, on the other. Manhattan has more telephone lines than exist in the total of three dozen sub-Saharan countries (French, 1998).

Singapore, China, Iran, and Vietnam want to limit uncontrolled access to the Net. Singapore, for example, is trying to prohibit its Internet providers from distributing "material that spreads 'permissiveness or promiscuity' or that 'depict(s) or propagate(s) sexual perversions such as homosexuality, lesbianism, and paedophilia" (Singapore Broadcasting Authority, 1996, quoted in Guy, n.d.). Ironically, the microtechnology on which the Net depends is produced in Third World countries, often by exploited labor.

The international telecommunications quadrennial meeting, Telecom 95, in Geneva adopted the slogan "Connect!" referring to the rapid eradication of boundaries that separate the telephone, computer, and media industries. Clearly, this was directed at only the top 50 developed nations, not the 48 poorest ones. The poorest nations have fallen further behind in the potentially liberating effects of telecommunications.

There are information "have-nots" in the United States as well. Blacks are far less likely to use the Net than are whites, and anybody who is poor is not likely to use it. The Department of Commerce notes that about 20% of America's poorest families do not have telephones, and only a fraction of those that have them can afford the computers and peripherals needed to participate in the information age. In the South Bronx in New York City, the problem is how to get poor people wired. One quarter (25%) of the participants in Lifeline, the local telephone company's program to provide basic phone service to families receiving some type of public assistance, live in the South Bronx, and less than 5% of all households there have computers. Even if more households could afford to buy computers, they lack the skills and training necessary to operate them. The New York Public Library has introduced LEO (Library Entrance Online) for local libraries that will allow a user to search the library's 50 million holdings and to explore the Net. But online time will be limited, and eventually the user will have to pay for printouts ("Information Democracy," 1995).

Considering the democratic performance of cyberspace, we also have to question the creation of both hypothetical communities and hypothetical identities on the Net. Elaborately constructed fantasy worlds created on the Net are known under a variety of acronyms such as MUD, MUSH, MOO, MUSE, and MUCK. The MU* (with the asterisk as a wild card) defines a multi-user, network-accessible user interface that is entirely textual (Guy, n.d.). These domains are highly interactive, computer-based extensions of fantasy role-playing games developed during the 1980s in which players invent fictional characters or *handles* and move these characters about as if they were puppets on strings. Often, actions are prefaced by comments such as `IC:` (in character) or `OOC:` (out of character) to stress a difference that is important to the play, even though it is difficult to ascertain the boundary that has been crossed. These fantasy games allow extravagant experimentation with the definition and self-invention of subjectivity, but there is an addictive quality to the games as well and a sense of dwelling in a metaphor without living the experience.

Mobile identities undermine the notion of authority and normative behavior. Because the Net is primarily a language-based realm, words can wound, and speech or language can constitute injury. As Nobel Prize-winning author Toni Morrison has said, "Oppressive language does more than represent violence; it is violence" (quoted in Butler, 1997, p. 6). The problem appears to lie in the metaphorical term *cyber-*

space. Considering the Net as space masks the fact that subjectivities on the Net are disassociated from their bodies or disembodied. Gender, race, and class become irrelevant and constructed at will. Hence, the tendency is to downplay responsibility and/or agency. As the Internet decreases our awareness of reality and increases privatization, a vacuum of responsibility arises in an online community. "Netiquette rules" are rudimentary.

In the "ecstasy of communication," obscenity begins. As Pamela Gilbert reports, aggressive behavior constituted "merely" of words is routinely displayed on the Net. There is a tendency to use aggressive words, to fight or flame at the least provocation, and to act out invasive or provoking behavior (Gilbert, 1996). An entire vocabulary has developed to describe this excitable speech. A *flame* is a message posted as an insult, and it usually elicits a response. The result is a *flame war,* a name-calling free-for-all of fighting words. A *troll* is a deliberately provocative post targeted to an audience the poster hopes to annoy; it is flame bait intended to provoke a war of words. On the other hand, an indiscriminant posting, aimed at the widest possible group, is called *spam* and generally sets up a cacophony, drowning out any hope of intelligent discourse (Guy, n.d.).

Stalking on the Net uses many of these harassing procedures but adds another dimension, as Gilbert describes. After she had terminated a relationship with a former colleague, he began to stalk her on the Net, violating her sense of personal space. He showed up on *listservs* where Gilbert was active. He searched her e-mail and obsessively traced her log-ons and whereabouts on the Net. She had a feeling of being watched every time she was posted or posted a message. There was nowhere on the Net she could safely go without being watched. Then, he threatened to send pornographic photographs that he had collected of her to her colleagues and to universities' search committees where she was seeking jobs. She felt as violated and unprotected on the Net as she would had she taken a walk late at night down some dark and dreary city street. The Net, like any physical space, has its noir side that is far from safe.

Online harassment appears to be a gatekeeping device that limits the extent to which women feel comfortable participating in cyberspace. Rowdy, group-flaming discourse deters women who tend to prefer semi-private e-mail discussions. With any new communication technology, there always is a backlash problem of allowing the wrong type of communication to take place. In the United States, a prime example is the proliferation of sexually explicit material available to children under 18 years of age, known as *cyberporn.* In response, the federally legislated Communication Decency Act was passed in 1996. This act has been deemed by the Supreme Court to be detrimental to the freedom of speech protected by the First Amendment of the U.S. Constitution. The question before the court rested on the decision of whether the Net was to be considered to be like radio and television or like the telephone and the newspaper. The former are so pervasive in the home that they require strict scrutiny and censorship, whereas the latter are considered to be discrete technologies and discourses not emitted from a central location. Hence, the latter constitute free speech that must be protected. Still, questions remain. Is cyberspace a closed-off space beyond the range of social norms and common decency? If not, then what type of regulation might help this electronic space to become truly democratic and open?

The Framed City Matrix

Let us look at the nodes and interstitial spaces of the matrix of cybercities. It appears, at least metaphorically, as if Michel Foucault's spaces of enclosure have been turned inside out in cyberspace. The marginals, however they are defined, are left outside of the protected zone of the shopping mall, the campus, and the walled community as well as outside the Internet, the credit-card/ATM system, and privileged trade nation agreements. These outsiders haunt and invade the interior. The marginal has become our postmodern monster, beyond the norm and increasingly invisible. This fear has caused an outpouring of walled cities in which the wall represents a boundary, a partition, a border that encloses and excludes much the way that cells of a matrix do.

Here, we can continue to deploy the analogy between the electronic matrix of cyberspace and the actual space of cities, but now we counterpose the purity, cleanliness, and synthetic quality of cyberspace—the visual matrix of space in the clean machine of the computer memory—in opposition to the dirt and seediness of real city space. The tendency is to introduce into real spaces of the city the synthetic purity of cyberspace, filling in only the nodes of the matrix. These imaginary assemblages and decorated places are the result of our taste for the artifice, simulation, and the computer-generated model.

Marvin Minsky, one of the early explorers of artificial intelligence, conceives of mental images as active structures or frames that help us to organize reality. In "A Framework for Representing Knowledge," Minsky (1975) writes,

> Whenever one encounters a new situation (or makes a substantial change in one's viewpoint), one selects from memory a substantial structure called a frame. This is a remembered framework to be adapted to fit reality by changing details as necessary. A frame is a data structure for representing a stereotyped situation, like being in a certain kind of room or going to a child's birthday party. Attached to each frame are several kinds of information. Some of this information is about how to use the frame. Some is about what one can expect to happen next. Some is about what to do if these expectations are not confirmed. . . .
>
> Collections of related frames are linked together into frame systems. The effects of important actions are mirrored by transformations between the frames of a system. . . .
>
> The frame systems are linked, in turn, by an information retrieval network. When a proposed frame cannot be made to fit reality—when we cannot find terminal assignments that suitably match its marker conditions—this network provides a replacement frame. (quoted in Bartels, 1993, p. 50; see also Johnson, 1995)

Minsky's frames revolve around memory in the sense that an individual selects a remembered framework. There also is a geographical analogy, a network of places that enables thought to move from one frame to another. One's behavior is modified by the fear of being outside a frame in a nontypical situation, so one's sense of reality is organized into preestablished frames making only small and gradual updates to the system. In a world of accelerated change, dislocations of meaning and failed correspondence with framed structures of knowledge are inevitable. Yet, the general tendency is for individuals to hold onto unrevised interpretive frames, no matter how discontinuous and divergent from reality they might be, until new frames of reference are accepted.

In an analogous manner, we find many frame-like zones, or rule-driven simulations or representations composed of structured correspondences, in our regional cities. Ada Louise Huxtable, in a recent article titled "Living With the Fake, and Liking It," calls these zones "the real, real space" of the city. They are surrogate experiences, synthetic settings, and artifices. In this blurring of boundaries, she argues, the spectator of city spaces cannot distinguish between the real and the false or tell the difference between the real fake and the fake fake (Huxtable, 1997).

The authentically unauthentic has become a new urban design frontier. "History repeats itself neither as farce nor as tragedy but as themed environment" (Huxtable, 1997). Huxtable bemoans the loss of "connoisseurship," that is, the training of the eye, the informing of taste by direct contact with the real thing and the true work of art. She specifies this contact as "the related sequence of close knowledge and informed taste by which works of art can be accurately understood, compared, defined, judged, and enjoyed. There is no replacement for this primary experience—the direct connection" (p. A40).

Why this nostalgic cry for contact with the real thing at the very moment when computer-generated simulation has confused our sense of authenticity? Are we actually afraid that the image of the city has been tampered with by simulated copies, or is this a cover-up for more fundamental fears?

Huxtable confuses the distinction between imitation and simulation. Imitation holds out the belief in an authentic origin, whereas simulation goes further. It generates the semblance of a nonexisting reality and often reveals the very apparatus of its own creation. Is not the projection of an image that we fear is false (i.e., not real or optically true such as a theme park image, the image of New York in Las Vegas or Disney World) a fear that we are losing control or mastery over the object world? In other words, in the edgy alliance between the actual and the virtual, the material and the immaterial, the imitated and the simulated, the projection of an image of the city that is believed to be false, that is not based in reality or optically true, questions our symbolic or imaginary possession of space (Kembler, 1996). To be more precise, as the status of the photographic image as documentary evidence is destabilized and its relation to memory and history is disassembled, interesting questions of epistemology, representation, and truth are raised about how we perceive and apprehend the world in which we live (Robins, 1991, p. 58).

Photographic documentation once served as a reality test. It was evidence that something actually happened, that reported events were true; photographic coverage of the Vietnam War and the pro-democracy movement in Tiananmen Square are two good examples. But the computer has had a major impact on photographic veracity. With electronic imaging, photographs no longer are sacrosanct from editorial modification. Digital manipulation of an image can add or subtract elements and enhance its lighting or color on command, so that its "truthfulness" or reliability as a record of reality becomes problematic. Computers offer the image maker the ability to transform, change, and move images. This same information system, with its mathematical models and algorithms, can generate a "realistic" image, reinventing the world from scratch. Although a computer image still might appear realistic, its referent is not. Reality no longer is being represented; instead, it is being modeled and mimicked. Whether computer manipulated or computer generated, electronic visualization constructs artificial domains and simulated environments. This image space appears to be completely autonomous, a surrogate reality that no longer can be judged for either its authenticity or its accuracy. Yet, we worry about the authenticity of urban space.

Returning to these artificial zones, image schemata, or frames, a major claim is made here that urban space is transformed in some manner by cybernetic, representational space. Because the matrix of cyberspace is a cognitive map (i.e., a mapping of conceptual space), the more features that are elaborated, the more the mind can navigate and negotiate the space inside its well-articulated nodes. Hence, the matrix of cyberspace becomes the way in which we accept and/or legitimate what has happened to cities as we no longer question an image of the city parceled into zones—business investment districts, common interest developments (CIDs), theme parks, malls, historic districts, special urban design districts, and the like. The move into cyberspace favors decentered, dispersed, and discontinuous sprawl, in contrast to self-contained zones. It allows for the interstitial areas of silence, estrangement, and decay to remain outside the grid. Cyberspace and city space go very much hand in hand here, enabling the viewer to forget the in-between, interruptive spaces. This is reiterated when the space of the city and the space of cyberspace are conceptualized as matrices.

The boundary between surveillance and city spaces breaks down with the production of fortified enclaves or walled cities as framed and privatized zones of the regional city. By the late 1980s, according to Evan McKenzie, approximately 40 million Americans lived in CIDs, that is, about one person in eight. CIDs also are called *gated communities* or *walled-in communities* and include cooperative apartments, condominiums, and new towns. They will be the predominant form of housing by the 21st century (Judd, 1995; McKenzie, 1994). Residents own or control common areas or shared amenities. An owner has to abide by a variety of covenants, contracts, and restrictions that may dictate the minimum or maximum age of residents, hours and frequency of visitors, color of paint on a house, style and color of draperies hung in windows, size of pets, number of children, parking rules, and patio and landscaping controls (Judd, 1995, p. 157). Because they impose reciprocal rights and obligations enforced by a private governing body, lawyers call CIDs "association-administered servitude regimes."

These are spatially segregated communities, and on a small scale, they simulate the functional order of the modernist city. Discrete areas are set aside for residences, work, leisure, and the necessary transportation corridors that link together these places. Look at the master-planned community of Irvine, California, on the southern edge of the regional sprawl of Los Angeles. The map of Irvine looks like a matrix, with its grid of highways and its series of well-designed nodes or theme parks, for Irvine is only 10 or 15 miles south of Disneyland.

The Irvine Company is in charge of developing this regional city and targets its image to two markets: businessmen working in the fields of advanced technology or luxury goods and residents with medium-high incomes who are firmly attached to the region's economy, are lovers of nature, are active in sports, and want to live in an extremely homogeneous social and economic environment ("Il Ranch di Irvine," 1996, pp. 76-77). In each of its residential areas, the values of nature, family, and work are bound together in "city Identity." Each residential nodal point is publicized as a tourist village in which everything is designed for a resident's well-being. The basic organizing unit of this new urbanism is the neighborhood—limited in area, revealing a well-defined edge, and structured around a defined center of activity. The optimal distance from center to edge is a quarter of a mile, and each neighborhood center contains shops and businesses and also might have a post office, a meeting hall, a day care center, and/or religious or cultural institutions. A fine network of streets and paths connects these neighborhood spaces.[3]

As the image of the city becomes important for high income-generating citizens in well-designed nodes of the city matrix, it simultaneously bears the marks of exclusion as it withdraws from the spaces left outside of its grid. Strategic control over this image is important, and conflict increases dramatically within the interstitial spaces of alterity. Containment takes a number of forms. One is the creation of social control districts in which criminal and civil code enforcement merge with land use controls (e.g., special anti-graffiti districts, anti–red light zones, new nuisance uses such as garage sales in Los Angeles, homeless districts in Miami, drug enforcement zones on the Lower East Side of New York). In these zones, the definition of *nuisance* has been extended from the health, safety, and general welfare issues that historically controlled the location of noxious industries and the placement of sweatshops and manufacturing uses to people's behavior, even though people cannot be fixed in place in the manner of land uses.

To protect the rights of the homeless in Miami, for example, a federal judge ordered the city to create "safe zones" where the homeless could sleep, eat, bathe, and cook without being arrested. The judge found that "Miami's policy of using ordinances on vagrancy, loitering, curfew, and disorderly conduct to bar the homeless people from public areas violated the constitutional provisions of due process, equal protection, and freedom from cruel and unusual punishment" (Rohter, 1992, p. A16). In the specialized zones that had been set aside, the police were expected to act with restraint. Lawyers for the homeless, however, believed that the homeless still would be arrested for sleeping or eating in public places that were not in one of the two designated safe zones; the homeless were categorized as eyesores, they smelled bad, and they discouraged investment and tourism. Some city officials also argued that safe zones should not exist; one cannot create areas, the mayor said, where certain people are immune from regulation.[4] Shutting out homeless people in the end is but the inverse of shutting ourselves up within protected domains of the framed city matrix because anything confused, unpredictable, and random (i.e., without an assignable place) threatens the analytical order.

Matrices and Maps

Returning to the analogy between matrices and the regional city, much of the theory of cyberspace assumes implicitly or states explicitly that a profound mutation has taken place that displaces the traditional Western space of geometry, work, the road, the building, and the machine. Instead, novel forms of diagramming, bar graphs, spreadsheets, matrices, and networks are expressive of "a new etherealization of geography" in which the principles of ordinary space and time have been altered beyond recognition. Michael Heim, in "The Erotic Ontology of Cyberspace," writes, "Filtered through the computer matrix, all reality becomes patterns of information. When reality becomes indistinguishable from information . . . , the computer culture interprets all knowable reality as transmissible information" (Heim, 1992, p. 65).

The pattern of information forming the architecture of computer memories is based on Boolean logic outlined by English logician George Boole. In Boole's (1850/1958) book, *An Investigation of the Laws of Thought,* he demonstrated how thinking follows clear patterns and mathematical rules (Hofstadter, 1995). Boolean logic has become the metaphor for the computer age; it shows how we interrogate information and how we scan through flows of data. This Boolean search guides our subconscious when we interact with the computer interface; this is how we begin to model the world as yes/no, on/off, and 0s/1s. In the digital arts, files are stacks of information, searched by keywords based on the Boolean algebra of AND (x and y both are present), OR (either x or y is present), and NOT x (x is not present). Furthermore, this binary logic is displayed graphically by Venn diagrams or images that classify things and then shuffle them into the appropriate sets of boxes. In the end, Boolean logic is the governing relationship that determines inclusion or exclusion.

Inevitably, direct experience declines, for the operations of Boolean logic are removed from subject matter and involve an abstract formal detachment having no inherent connection to things we directly perceive and experience. Heim (1993) notes,

> Boolean search logic cuts off the peripheral vision of the mind's eye. The computer interface can act like the artificial lens that helps us persist in our preconceptions. . . . We may see more and see it more sharply, but the clarity will not hold the rich depth of natural vision. The world of thought we see will be flattened by an abstract remoteness, and the mind's eye, through its straining, will see a thin, flattened world with less light and brightness. (p. 25)

Cyberspace devotees assume that they are jacked into this binary matrix of information flow. Yet, they make a rhetorical jump, crossing boundaries from a technology based on binary logic to the fictional realm of a seamless human-machine symbiosis. These boundary crossings are based on the belief that the world can be understood by mathematics. They constitute an idealized projection of an underlying order whose mathematical substructure reveals the coherence not of a material world but rather of an objective realm of ideal forms (Markley, 1996, p. 67). In other words, cyberspace is a complex idealization. It offers the alluring fiction of limitless possibilities and connections rather than a stale recasting of oppositional binary logic, but its principles of self-organization remain mathematical and Boolean (p. 69). Whatever way in which the denial is posed, cyberspace is based on binary structures that characterize traditional views of Western thought—the mind-body and idea-matter oppositions that have been around for centuries, yet oppositions that cyberspace advocates seek to overcome.

The problem is that the real world is blurry and logic is incapable of dealing with new and unanticipated types of situations. Nor does logic deal with pattern recognition or activities that involve visual perception, abstraction, and comparison. In Boolean set theory, an element is either in or out of a set (0/1), but in reality, there are fuzzy sets in which things lie neither in nor out but rather somewhere in the middle. To include fuzzy thinking, connectionist models of cognition or associative memory and image-based modes of thought need to be examined.

Assemblages represent a form of associative thinking. Created by the liberal use of photocopy machines, silkscreen printing, and instant-developing cameras, an assemblage suggests that categorization is based on human experience and imagination. Thus, associative grammars are erected on strategies of analogy and circumlocution that facilitate unexpected juxtapositions and overlapping images. Associative thinking is neither linear, progressive, rational, nor conclusive. It is engendering and utilizing, and it involves recursive reflexivity, loops, and returns. Furthermore, associative memory is based on storing a given piece of information next to similar information, not in the standard pigeonholes of the library grid or a tree-structured encyclopedic recall. The latter are networks based on Boolean logic or matrix mathematics (Caudill, 1992, pp. 73-88). Rather than focusing on nodal points and linkages in the matrix, it is the oscillation between the contents of the nodes that becomes the concern in associative grammars.

How does associative thinking relate to mapping or imagining the physical form of regional cities where matrix mathematics seems firmly in control? As discussed previously, Deleuze and Guattari (1987) proposed an alternative to the conceptual tools we normally use to order and hierarchialize the actual territory of the city. Their virtual postmodern map

> is open and connectable in all of its dimensions; it is detachable, reversible, susceptible to constant modification. It can be torn, reversed,

> adapted to any kind of mounting, reworked by an individual, group, or social formation. It can be drawn on a wall, conceived of as a work of art, constructed as a political action or as a meditation. . . . it always has multiple entryways. (p. 12)

Indeed, this ideal of connectivity and malleability produces unpredictable conjunctions capable of bonding virtually anything. It holds out the alluring fiction of limitless possibilities and endless connections.[5] It is part of the fiction of cyberspace that participants exist in " 'dynamic interaction with information' and consequently 'space (and experience) are pervasive rather than dualistic' . . . 'both/and inclusions rather than either/or dichotomies' " (Markley, 1996, p. 65).

Thus, electronically mediated existence or the virtual reality of cyberspace becomes a means to imagine infinite productivity, the limitless possible metaphors of information and exchange (Markley, 1996, p. 75). The problem is the conjunctive hinge between virtual reality and, for want of a better word, actual reality. Cyberspace has no way of treating boundaries, no way of crossing from virtual into actual reality. This problem exists with any mathematicization of space and time. It always reduces real space and time to the formal applicability of its conceptual tools. A virtual reality machine has to process vast amounts of data and feedback responses, and simplified processes have to be employed to reduce the complexity of reality to manageable form (p. 59).

The argument being developed here distinguishes between different computational models of cognition and relates this to how we map or imagine the form of a regional city. The conventional model on which computers currently are based takes cognition to be the manipulation of symbols in accordance with preexisting computational rules. Kevin Lynch's image of the city is based on such symbol-manipulating procedures of cognitive mapping (Lynch, 1960). On the other hand, the connectionist or associative network model operates as a parallel processor and has yet to be materialized as the architecture of a computer memory. This cognitive map more adequately expresses the complexities and nonlinear dynamics of regional cities. The latter model figures cognition as the spread of activities across a network of interconnected units. The connections between these units, rather than the units per se, take on the pivotal role in the functioning of the network (Wilson, 1966). Traditional theories of cognitive functions assume that information flows in a linear and sequential manner. Connectionist models move away from causal linearity toward a web of interconnections. Thus, cognitive processes are supposed to be distributed and parallel, not sequential and linear.

Von Neumann's machine, the grandfather of computation theories, was based on a series of rule-governing processes that literally were written out in propositional logic and stored inside a computer program. They established a hardware-software distinction in which the software determined cognitive functions and the hardware was simply a machine on which the program was implemented. Connectionist models reject this software-hardware distinction. There is no central processing unit and no stored rules. Cognitive functioning is controlled by the difference in weights between units, the general wiring of units, and a series of learning rules. Knowledge no longer is stored in locatable pigeonholes of an electronic memory, the nodal points of the matrix, but rather is stored in the links/connections, the interstitial spaces between units where memory becomes the effect of differences between units and across the network (Wilson, 1966).

Perhaps we should turn to Italo Calvino's *Invisible Cities* for a demonstration of associative memory. He argues against a fixed spatial form (such as the matrix) being used as a cognitive map of the city. Calvino (1974) notes that the computer provides a theoretical model for the most complex processes of memory, mental associations and imagination. Being based on rational thought, it simultaneously opens up toward multiple routes and unbounded wanderings. Finally realizing the *ars combinatoria* proposed by medieval monk Raymon Lull, electronic brains make instantaneous calculations of unthinkable complexity, a "triumph of discontinuity, divisibility, and combination over all that is flux, or a series of minute nuances following one upon the other" (Calvino, 1986, p. 9).

Calvino's mathematical structure spreads his tales of 55 cities over 9 sections and classifies them into 11 sets of city types (e.g., "cities and memory," "cities and desire," "cities and names"). A rotating device determines that after the fifth appearance of a tale from a given city set, that type is retired and another one is born ("cities of memory" is finished and "trading cities" begins). Because the referent of these cities is invisible, their meaning must appear on the surface as signs change places and regroup. The reader must focus on the literal reading of the arrangement of tales in spite of the fact that the abstract nature of this serial play tends toward accumulation and incompleteness (James, 1986).

Poor Kublai Khan desires a location map to link these cities together, a map he can master so that he can control all of this space. Yet, he is confined to follow the author's open-ended process that moves in indeterminate directions. Thus, the author declares, "I will put together, piece by piece, the perfect city, made of fragments mixed with the rest, of instants separated by intervals, of signals one sends out, not knowing who receives them" (Calvino, 1974, p. 164).

Writing on the Net

Finally, let us turn to the establishment of a textual virtual environment on the Net. As one Internet site advises the user who enters its space, "You are your words," underscoring the fact that the Net is predominantly a territory circumscribed by sentences, phrases, and words skittering across cyberspace in different chat rooms, homepages, and electronic mail (Bennahum, 1997, p. 23). Your body enters this "space" as a form of writing; it literally becomes a cybertextual body. Because language wraps the Internet, the entire environment becomes a textual projection.

In this word space of electronic culture, a work of art (be it verbal or visual) undergoes perpetual transformations and desubstantiations. Indeed, electronic writing is deeply playful; it focuses attention on the rhetorical surface, an art intended to persuade. Its game of play depends on locating attention and emphasis through its use of topics, preformed arguments, phrases, and discrete chunks of verbal boilerplate that can be cut, pasted, and repeated at will (Lanham, 1993). The conventions of reading from left to right and from top to bottom are challenged by diverting these norms and instead rotating lines, using inverse mirroring of words and sentences, and ignoring the familiar grid of the page (Dworkin, 1996, p. 391). Once again, cyberspace pushes us to think beyond the matrix or the grid that has dominated Western conceptualization since the beginning of writing.

Look at several pages of the Times Square Business Improvement District booklet, designed by Two Twelve Associates, and examine the parts of the message that are marked and those that are not:

TIMES SQUARE
is as much an
(idea)
as it is
a place.

"The slightly Eccentric,
MAGNIFICENTLY artful
EXCITEMENT
and thrilling variety of . . .
TIMES SQUARE."

The first text, by shifting type and altering scale, not only directs attention but also plays with the relations of greater and lesser than; it presents the amusing thought that Times Square is an idea first and a place second. The second text, by shifting typefaces, styles, sizes, positions, turns the simple statement into a series of zoned and ranked activities. These zones are linked to the actual space of Times Square that is full of excitement, certainly magnificent, possibly eccentric, and slightly artful. Placing the emphasis on rhetoric more than on syntax, this form of writing allocates emphasis and directs attention in a nonlinear system of statements aimed at describing the confusion of everyday life in the public space of Times Square (Drucker, 1994).

The gestural language that encompasses the advertising billboard seems to swirl around the proclamations of Venturi, Scott-Brown, and Izenour (1972) in *Learning From Las Vegas.* Back in 1968, they acknowledged the rhetorical surface of the cityscape, noting that the successful text in the commercialized culture of the late 20th century was the one that combined high-speed communication with maximum information (Perloff, 1991, p. 93).

Whereas *Learning From Las Vegas* was concerned with architecture in the age of television and the automobile, Venturi has now turned his attention to new modes of electronic communication. Venturi's (1996) latest book, *Iconography and Electronics,* is primarily concerned with language of the direct address and the manner in which statements communicate. Architecture, Venturi proclaims, involves shelter and symbol; it "embrace[s] signs, reference, representation, iconography, scenography, and trompe l'oeil as its valid dimensions" (p. 3). But rather than examining the signs of Las Vegas and exploring the iconography of the commercial strip as once was the focus, Venturi now is intent on studying the electronic images displayed on Jumbotrons and light-emitting diode (LED) mosaics that emit their messages day and night. The Las Vegas Strip reveals dramatic changes since 1968; it has been urbanized, with its linear strip settlement recomposed in superblocks, its parking lots either filled in with large hotels or converted into front lawns, and its traffic so congested that its streets soon will be pedestrianized. Signs have evolved into scenography and neon replaced by LED matrices of ever changing images and graphics (pp. 123-128).

Venturi's underlying question is clear: What will be the future relation of electronic technology to writing, architecture, and urban space if the former transposes the latter? The answer requires renegotiating the entire word-image (or alphabet-icon) relationship. This revolution in the material form of writing affects architecture and urban space in that they become the public surface on which electronic texts are displayed. It offers new forms of expressivity, new relationships to the word and to typography, and new intellectual techniques of cognition. Electronic writing searches for ways in which to bring comedy and play into the seriousness of print and design; it aims to make communication interactive, not reliant on conventional or canonical expressions.

Lessons can be learned from Times Square/42nd Street because this public space always has been famous for its electronic one-liners; the strip zipper blurting out news bites on the Times Building was put in place in 1924. Over the past two decades, textual artists such as Jenny Holzer have been preoccupied with these one-liners, what language poets call *aphorisms.* Her "truisms" (1977-1979) were a series of "mock cliches" or throwaway paste-up texts such as those she found scattered about Times Square warning men to stay away from the vice in the area, exclaiming how they would get leprosy if they overstepped a magical circle drawn around the place. She was amazed at how the word *leprosy* on a poster stopped her short. This led her to experiment with her own short texts and to place them in public space wherever she wanted.

At first, Holzer did not exploit the expressive potential of various typefaces or construct her texts in zones of activity. Her aphorisms were stereotypical one-liners typed on white paper in black ink. Alphabetically arranged line by line, her truisms crossed over the border from literary to visual art and raised many questions in their wake. How do we read the following lists? "Abuse of Power Comes as No Surprise" . . . "Action Causes More Trouble Than Thought" . . . "Alienation Produces Eccentrics or Revolutionaries" Are they a manual for operating in contemporary times? Are these alphabetical word spaces to be compared to other verbal orderings of experience such as lists, associations, fragments, those "structures of accretion" that pile up disparate things and only slowly begin to make contradictory and complicated sense? Are we browsing through catalogs—flipping, scanning, scrolling, or linking through mobile electronic texts—as the eye and the mind catch similar or dissimilar meanings and thoughts? They seem to be a form of nonlinear associative thinking where meaning is intended to slip in and out of focus. Meant to provoke and to prod the spectator to stop and to read, in the end, these aphorisms problematize communication and stand on the edge of chaos.

Venturi's writing also has been involved with one-liners and uses forms of direct address. In *Complexity and Contradiction in Architecture,* Venturi (1966) told the reader that "less is a bore" and that "I like elements which are hybrid rather than 'pure,' compromising rather than 'straightforward,' ambiguous rather than 'articulated,' . . . inconsistent and equivocal rather than direct and clear" (p. 16). By stressing the words he dislikes—pure, straightforward, articulated—Venturi highlights his own double-coded preferences for mundane objects stripped of all pretensions yet wildly productive of a multivalent, multivocal architecture. Such was his rhetorical intent—to bring the noisy materiality of the outside world of the street and the billboard inside the clean and distilled realm of modern architecture so as to shatter its tranquility like an explosive device.

Venturi's attack continues in *Iconography and Electronics,* where he opens on a "sweet and sour" note. There is particular sweetness in the information age where electronic signage transforms words into evanescent, instantly transformable signs and opens onto a variety of languages, both vulgar and tasteful. But there are sour notes as well, a tone that dominates his one-liners presented as a list of "Mals Mots: Aphorisms—Sweet and Sour—by an Anti-Hero Architect" (Venturi, 1996, pp. 299-329). Venturi develops a "flaming" critique of the *retardataire avant-garde American architects* who have transformed architecture into a theorized and immaterial form. His rhetorical play enables formal pleasure to balance conceptual thought and self-conscious stylistics to alleviate theoretical despair (pp. 2-16). Hearing echoes of Holzer, for it is difficult not to see these statements flashing across a spectacolor board and flung down at the reader as so many "truths," Venturi uses forms of public address to express personal feelings, ambivalences, and even insecurities, thereby mimicking the self-conscious manner of electronic space.

Venturi offers a cluster of volatile statements titled "Valid Rantings" that displays posturing, suggestibility, and play:

> "I am always out of step" and "I am an exhibitionist: I go around exposing my thoughts" . . . "It is better to be good than in—I think" . . . "Challenge me, don't harass me!" . . . "Am I being sour?— Yes, but also sweet as that combination works to illuminate." "It's OK to be cranky if you're perky too" . . . "Am I the marginal nerd?" "Remember, my positivity complements my negativity and, I hope, embraces wit."

Eventually, Venturi (1996) reaches his goal and presents "Iconography and Technology":

> "Viva iconography *and* scenography in architecture." "Can we learn from the vivid art of advertising of today?" . . . "Iconography is all over [T-]shirts—why not buildings?" . . . "Billboards perversely represent the civic art of today—to be cherished 100 years from now as significant elements for historical preservation; Houston will be for billboards what Williamsburg is for colonial." . . . "Today's fear of electronic technology for evolving iconographic surfaces in architecture resembles yesteryear's fear of engineering technology for a machine aesthetic." . . . And now, from decorated shed to virtual box." . . . "Light is our essential material—more than bricks (and even wire frame guy wires)." "Light is our essential medium—not as 'veiled' and 'luminous' planes that are electric but as vivid iconographic decor that can be electronic." . . . "Virtual variety." . . . "Iconography as graffiti glorified." (pp. 325-328)

This brings us back to the city and, in particular, to urban spaces such as Times Square that are filled with writings on the wall. Zooming in and out of an image, reversing the traditional figure-ground relationships, or playing with ornamental/purposive features, electronic writing allows the writer to shift scales at will. The viewer/reader looks at writing as a material surface or a series of pixel patterns. Just as Roy Lichtenstein's 1960s comic book paintings forced the viewer to look at the surface pattern of dots as a design motif, Venturi envisions the gigantic blown-up surface of electronic writing as the nonlinear public space into which we enter. Writing becomes the interactive and volatile iconographic surface we must learn to playfully traverse and enjoy.

Conclusion

The analogy between the computer matrix and regional cities, or the spatial orderings that cybercities erect, asks us to make a qualitative leap from virtual to physical space, erroneously assuming that the boundaries separating these spaces can be crossed with ease. Behind all the hyperbole surrounding cyberspace and the Internet, challenges must be faced—about the future of democratic public space (whether virtual or real) and about increasing privatization, commercialization, and hierarchical control that create a new periphery (those locked out of the zones of cyberspace or those left behind in the digital divide between the well-defined frames and zones of cybercities). The problematic construction of subjectivity in computer-mediated communications needs to be scrutinized. There also are challenges to face concerning the basic assumptions of mathematics that sustain both the architecture and language that control events in cyberspace. In the interface between writing and architecture, we need to keep in mind both the materiality of space and its representational images as they oscillate back and forth in the construction of meaning and as they open us to new explorations and playful reformulations.

The hybrid condition of cybercities challenges the long-held privileged status of Cartesian geometry, the map, and the matrix or grid. Infrastructural links and connectors as well as information exchanges and thresholds become the dominant metaphors to examine the boundless extension of regional cities. We need to look at the space of the city in a different manner to envision the thresholds and loops it now represents. A map with its pictorial icons and static spatial relationships cannot represent the nonlinear and interactive flows of cybercities.

On the edge of chaos, the contemporary city might no longer be readable in the manner that Lynch (1960) prescribed in *The Image of the City.* Yet, there always will remain the need for some conceptual filter to make sense of the confusion of turbulent urban conditions. Standing on the threshold between fixity and relativity, the natural and the artifice, the real and the imaginary, the liminal space of cybercities still waits to be imagined.

NOTES

1. SAIC's board of directors consists of retired defense and intelligence officials. See Guy (n.d.).

2. Review by Steve Cisler (sac@gala.apple.com) of Doheny-Farina (1996).

3. See Duany and Plater-Zyberk (1994). For another example, look at the Regional Grid Plan for Madrid: 1997-2017 (Comunidad de Madrid, 1996). Here, the radio-centric model of a spread city or the concentric ring theory of circumferential highways has been replaced by an orthogonal mesh as the orienting paradigm for the region's structural growth. "The units of this grid, network, mesh, or reticulum, represented on an analogue plan . . . , offer a new vision of regional land" (p. 12). The grid plan is based on the notion of sustainable growth throughout the entire region and is thought to be able to transfer the pressure of growth toward new adjacent spaces as opposed to hypertrophy and congestion implicit in earlier development models.

4. There have been similar ordinances on vagrancy, begging, and park curfews related to homelessness in Atlanta, Chicago, Las Vegas, and San Francisco. Chicago swept all of the homeless from its airport, and New York has a state law prohibiting begging in the streets (a case that advocates for the homeless have declared to be unconstitutional on the grounds that begging is a form of speech and, therefore, cannot be barred).

5. For a full definition of the conjunctive "and," see Doel (1996).

REFERENCES

Alexander, C. (1965, April). A city is not a tree. *Architectural Forum,* pp. 58-62.

Bartels, K. (1993). The box of digital images: The world as a computer theater. *Diogenes, 41*(3), 50.

Bennahum, D. S. (1997, February 16). I got e-mail from Bill. *The New York Times Book Review,* p. 23.

Boole, G. (1958). *An investigation of the laws of thought on which are founded the mathematical theories of logic and probabilities.* New York: Dover. (Originally published in 1850)

Boyer, M. C. (1996). *CyberCities: Visual perception in the age of electronic communication.* Princeton, NJ: Princeton University Press.

Butler, J. (1997). *Excitable speech.* New York: Routledge.

Calvino, I. (1974). *Invisible cities* (W. Weaver, Trans.). New York: Harcourt Brace Jovanovich.

Calvino, I. (1986). *The uses of literature.* New York: Harcourt Brace Jovanovich.

Caudill, M. (1992. *In our own image: Building an artificial person.* New York: Oxford University Press.

Communidad de Madrid. (1996). *Land strategy regional plan basis 1996.* Madrid: Direccion General de Urbanismo y Planificion.

Deleuze, G., & Guattari, F. (1987). *A thousand plateaus* (B. Massumi, Trans.). Minneapolis: University of Minnesota Press.

Doel, M. (1996). A hundred thousand lines of light. *Society & Space, 14,* 421-440.

Doheny-Farina, S. (1996). *The wired neighborhood.* New Haven, CT: Yale University Press.

Drucker, J. (1994). *The visible word: Experimental typography and modern art, 1909-1923.* Chicago: University of Chicago Press.
Duany, A., & Plater-Zyberk, E. (1994). The neighborhood, the district and the corridor. In P. Katz (Ed.), *The new urbanism* (pp. xvii-xx). New York: McGraw-Hill.
Dworkin, C. D. (1996). Waging political babble: Susan Howe's visual prosody and the politics of noise. *Word & Image, 12*(4), 389-405.
Fitting, P. (1991). The lessons of cyberpunk. In C. Penley & A. Ross (Eds.), *Technoculture* (pp. 295-315). Minneapolis: University of Minnesota Press.
French, H. W. (1998, January 26). In Africa, reality of technology falls short. *The New York Times,* pp. D1, D10.
Gibson, W. (1984). *Neuromancer.* New York: Ace Books.
Gilbert, P. (1996). On sex, cyberspace and being stalked. *Women & Performance, 9*(1), 125-149.
Guy, N. K. (n.d.). *Community networks: Building real communities in a virtual space?* Available on Internet: http://www.vcn.bc.ca/people/nkg/ma-thesis/title.html
Harmon, A. (1998, January 26). Internet group challenges U.S. over Web address. *The New York Times,* p. D5.
Heim, M. (1992). The erotic ontology of cyberspace. In M. Benedikt (Ed.), *Cyberspace: First steps* (pp. 59-80). Cambridge, MA: MIT Press.
Heim, M. (1993). *The metaphysics of virtual reality.* New York: Oxford University Press.
Hofstadter, D. R. (1995). On seeing A's and seeing As. *Stanford Humanities Review, 4*(2), 109-121.
Huxtable, A. L. (1997, March 30). Living with the fake, and liking it. *The New York Times,* pp. 1, 40.
Il ranch di Irvine a Orange County: La città che non imita la città [Irvine Ranch of Orange County: The city that does not imitate the city]. (1996). *Lotus, 89,* 52-99.
Information democracy [editorial]. (1995, November 20). *The New York Times,* p. A14.
James, C. P. (1986). The fragmentation of allegory in Calvino's *Invisible Cities. Review of Contemporary Fiction, 6*(2), 88-94.
Johnson, M. (1995). Swamped by the updates: Expert systems, demioclasm and apeironic education. *Stanford Humanities Review, 4*(2), 86-87.
Judd, D. (1995). The rise of the new walled cities. In H. Liggett & D. Perry (Eds.), *Spatial practices* (pp. 144-166). Thousand Oaks, CA: Sage.
Kembler, S. (1996). The shadow of the object: Photography and realism. *Textual Practice, 10,* 145-163.
Landler, M. (1995, October 8). Haves and have-nots revisited. *The New York Times.*
Lanham, R. A. (1993). *The electronic word democracy.* Chicago: University of Chicago Press.
Lynch, K. (1960). *The image of the city.* Cambridge, MA: MIT Press.
Markley, R. (1996). Boundaries, mathematics, alienation and the metaphysics of cyberspace. In R. Markley (Ed.), *Virtual realities and their discontent* (pp. 55-77). Baltimore, MD: Johns Hopkins University Press.
McChesney, R. W. (1996). The Internet and U.S. communication policy-making in historical and critical perspective. *Journal of Communication, 46*(1), 98-124.
McKenzie, E. (1994). *Privatopia.* New Haven, CT: Yale University Press.
Minsky, M. (1975). A framework for representing knowledge. In *The psychology of computer vision* (pp. 212-213). New York: McGraw-Hill.

Morris, M., & Ogan, C. (1996). The Internet as mass medium. *Journal of Communication, 46*(1), 39-50.

Moss, M. L., & Townsend, A. (1996, September 11). *Leaders and losers on the Internet.* Available on Internet: http://www.nyu.edu/urban/research/internet/interent.html

Ostwald, M. J. (1997). Structuring virtual urban space: Arborescent schemes. In P. Droege (Ed.), *Intelligent environments* (pp. 451-484). Amsterdam: Elsevier.

Perloff, M. (1991). *Radical artifice: Writing poetry in the age of media.* Chicago: University of Chicago Press.

Robins, K. (1991). Into the image: Visual technology and vision cultures. In P. Wombell (Ed.), *PhotoVideo* (pp. 52-77). London: Rivers Oram.

Rohter, L. (1992, November 18). Miami ordered to create homeless zones. *The New York Times,* p. A16.

Sterling, B. (1989). *Islands in the Net.* New York: Ace Books.

Venturi, R. (1966). *Complexity and contradiction in architecture* (Papers on Architecture). New York: Museum of Modern Art.

Venturi, R. (1996). *Iconography and electronics: Upon a generic architecture.* Cambridge, MA: MIT Press.

Venturi, R., Scott-Brown, D., & Izenour, S. (1972). *Learning from Las Vegas.* Cambridge, MA: MIT Press.

Wilson, E. A. (1966). Projects for a scientific psychology: Freud, Derrida, and the connectionist theories of cognition. *Differences, 8*(3), 21-52.

4

The Post-City Challenge

THIERRY PAQUOT

Have we not said everything there is to say about "the city" and its historical destiny? Library shelves are crammed with reports, theses, papers, essays, and other writings describing the city's ills and suggesting, although more seldom, a host of solutions. I am a conscientious reader of this type of literature (or at least of a tiny fraction of it, mostly in French) and a convinced, if not militant, city dweller, and I have observed that analyses of the assets and drawbacks of urban agglomerations usually harp on the same themes; imaginations flounder when trying to envision the future, or at least the future development, of this "object" we still call the city. On the one hand, there are those who wallow in nostalgia for the "good old days" (?) when cities were close-knit and compact. On the other hand, the experts muse not about the end of the city but rather about its growth and increasing encroachment on the countryside.

Stepping back to contemplate the planetary nature of current transformations, one wonders whether we are dealing with a change of scale or witnessing something completely different. This is where opinions vary, a reassuring fact. Cities, like life itself, never are all one thing or all another thing. They are full of color, with a variety of shades and contrasts. Perhaps we can say that the city, after occupying the center stage for several centuries, finally has been kicked aside by the history of humanity and upstaged by a new urban entity whose nature has yet to be understood. Thus, we are unable to acknowledge it.

Such is the theory I would like to present in this chapter. First, the city is now out of date. True, it has had its moments of glory—and of

EDITOR'S NOTE: This chapter was ably translated from French by Zoe Andreyev.

despair as well. I am thinking of cities-at-war, killing, sacking, burning cities; cities that have lost their heads and try to impose control through segregation, separation, and division; cities that have yielded to old totalitarian phantoms. Second, the urban entity is not a blank page; it has its own prehistory, its own past that determines its future, if only partly. Yet, so far, this vision has been purely static; a dash of time and movement would breathe life into it. In addition, we must include the elements of context—the geographic, demographic, climatic, cultural, technological, and other contingencies that represent more than just decor. Their role is essential in defining our two protagonists: the city and the urban entity.

Because we have begun in this vein, why not resort to fiction to explain reality? Never mind the plot; the important thing is how the characters perceive their environment. In Jean Echenoz's novel, *Un an,* the heroine leaves the dense city (the Montparnasse quarter in Paris) for Biarritz, a provincial town, and ends up wandering in a *no-man's-land* that is nevertheless urban. She survives in the city's interstices, on the edge of the "legal" city, a living demonstration of the fact that social exclusion is part and parcel of the social "machine" and that we must analyze both the mechanisms that create exclusion and those that promote inclusion (Echenoz, 1997).

The characters depicted by Douglas Coupland in the novel *Microserfs* are totally engrossed in their jobs in the computer industry. As a result, the urban background is obliterated by their working environment. Their work is extremely demanding and swallows up all of their energy, day and night. Their language is software-speak. Their view of the world revolves around computers. Their city, if such a word has any meaning for them, is limited to the places that necessity drives them to haunt—the house in which they all live together, the minimarkets where they buy food, their pals (if they have any), houses or apartments, the sports club where they work out to keep fit (because vitamin-enriched fruit juices do not suffice), roads and freeways. The geographic limits of this city are blurred, imprecise. The *virtual* world in which the characters live is real (Coupland, 1996).

The same can be said of many characters in recent literature. The space in which they move is extended, elastic, and sometimes even invisible. Time and speed are more important than their concrete location in space (Virilio, 1995, 1996). Personal relationships are determined by mobility, which depends on transportation and communication technologies. Mobile phones have considerably changed the plots of detective novels and movies; from now on, anywhere always will be somewhere.

Nevertheless, to be meaningful and play a role in the urban story, the terms *border, frontier, outside,* and *inside* must be able to call up or create their own specific set of mental references. A place exists only if it acts as a link, only if it is a place where something happens. Beyond such places lie the out-places (*hors-lieu*) that are out of the picture (*hors-jeu*) yet must be distinguished from the non-place (*non-lieu*), that is, the space that links us to others but where they themselves are not to be found, for example, the space of television (Auge, 1992; Younès & Mangematin, 1997).[1]

Searching desperately for *present* traces of the city is like thinking that one becomes a fish by diving into an aquarium; the very *nature* of the city no longer can be separated from the notion of "the urban" that envelops, penetrates, and transforms it. A city locked in by the urban can at best appear to be *what it used to be,* just as the heritage movement makes museums of what it seeks to safeguard. Cities *as they used to be* are not saved by tourism, although that is what one is led to believe. Every small town has *something* to preserve, restore, show, list in a guidebook, and include in a sightseeing tour. But what sight does one actually behold? More often than not, it is a brand new, reconstructed, video-monitored building surrounded by parking lots for which some type of past, part of the universal town mythology, has been refurbished. Therein lies the error. Each city is unique, and its history is one among many geographic, cultural, religious, linguistic, and ethnic histories.

How do we define the city's forms? Understand its meanings? Describe its mechanisms? The words *city, suburb, banlieue,* and *exurban* no longer mean much; the social and cultural realities to which they refer no longer are connected with actual physical locations. The suburban shopping mall functions as a real center, paradoxically turning the former city center into something peripheral. The same can be said for everything that has to do with the places of daily mobility—neighborhood, housing, public areas. These terms tell much more than their simple definitions would lead us to suspect. *Neighborhood* calls up the intricate web of itineraries weaving the thread of time into our daily lives. It does not necessarily imply neighborly relationships for the simple reason that one can be "close" while being distant in terms of space. Thanks to the phone or the fax, I am close to someone living many miles away. I can get in touch with him or her as easily as I can knock on my neighbor's door.

As a result of this distorted notion of proximity or distance, we have a surprising relationship between time and place. This relationship, called *chronotopia,* is a combination of the analysis of paces (rhythm-analysis) and that of places (topo-analysis) and provides us with an accurate idea of how we inhabit the real and imaginary world in which our bodies, dreams, fears, secrets, and certainties unfold. This is a world that more readily accepts our presence *among* others than *with* others. This is an essential distinction, and it represents the main challenge of the "post-city" era.

The post-city era does not follow chronologically the era of the city, nor is it the spatial continuation of the city into what can less and less be termed the *countryside.* Indeed, the post-city era is the era of universal urbanization. It represents a state of civilization more than a historical period or a form of land occupation. This post-city era has its own spatial forms and time scales; it accepts all sorts of configurations and creates new ones.

To give an example, seen from Europe, American cities come in two versions. The first, the European city, was imported by the colonizers. The second is that of westward-moving pioneers, cities such as Los Angeles, crisscrossed by highways. America also is the country of edge cities and gated communities (ranging from small to large). Urban "models" no longer are clearly circumscribed in the mapped world, and observing our surroundings does little to help us to identify any type of urbanistic-architectural truth. Strange combinations, the conventionalities of an epoch, the style of a social class, local architecture, outlandish prototypes, hidden details we can uncover, formless forms best forgotten—everything is possible in our urbanized world. Everything and its opposite. In particular, one cannot deny the responsibility of builders (perhaps representing the conscience of history—the moral of the story). Although urban civilization, which little by little has become our habitat, does rather gleefully collect fakes and trompe l'oeil decor, we can hardly be satisfied. For urban civilization to last, it must take by surprise, be tolerant, and accept diversity, even if it means running the inevitable risk of becoming indifferent.

In other words, urban civilization is banking on the evolution of singularity into plurality. But to understand the wonderful opportunity that such a civilization represents, we first must go back and examine the notion of city. The city is not easily described and analyzed; all of its multiple components have very different histories.

After all, what do we really mean when we talk about the city? The definition of the city as a very large number of people living together in a small, clearly circumscribed area of land is not very satisfying.[2] A city, as we might enjoy contemplating on nostalgic evenings, looks like a Renaissance depiction of Jerusalem, a cluster of houses with graceful gables, proud windows, and pointed roofs, huddling within large stone walls. One can almost feel the atmosphere, hear the shouts of the market vendors, imagine oneself slipping out the door of a house into a side street where happy children play. A beautiful dream city. It is ageless. Its only wrinkles are the graffiti on the walls and the cracks in the floors and ceilings that draw a map of fantasy in the atlases of our imaginations. Such a city does not exist. There is not *one* city; it always and necessarily is multiple. Historians speak of underlying urban structures, geographers speak of urban systems, and sociologists establish an opposition between urban and rural or oppose the city to other cities governed by different economic and political constraints imposed by the international and spatial divisions of labor. Still others—philosophers, poets, novelists, filmmakers—depict the city with caution, almost as if they were afraid of revealing a secret, of lifting a veil, of breaking wild Ariadne's fragile thread that guides them through an ever more complex labyrinth. For the philosopher, the city in the era of urban civilization is more than a stage set for an artificial and uninteresting play. It is the very stuff of which the urban being is made.

Urban Scales

According to Fernand Braudel, the city is "an anomaly of human settlement." In his book *Civilisation matérielle et capitalisme,* the chapter devoted to cities begins in this way: "Cities are like electrical transformers: they increase tension, accelerate exchanges, are endlessly churning human lives" (Braudel, 1967, p. 369). His vision of the city as a *fait de civilisation* is dynamic. All cities are different. Establishing a strict typology would be risky business. Nevertheless, we can observe common traits in Eastern and Western cities, in the cities of the past and those we live in today. In cities, money circulates in abundance, and these constant exchanges generate common attitudes, values, and practices that all city dwellers must adopt. States often originate in cities, and their aim often is to control, beautify, and populate cities as befits their growing prestige, wealth, and power. Capitalism weaves its global spider web, absorbing one city after another. That is about all that cities have in common.

Paris's Forum des Halles, the Roman forum, and the Greek agora are absolutely not related. The history of cities is not linear. To admire the ruins of ancient cities, one can visit archeological sites. However, a city's memory is to be found not in its ruins but rather in its words. Memory feeds on words like fire feeds on wood. A site without words, deprived of speech, ceases to be urban.

Yet, memory is unreliable. Heavy bronze plaques recounting local historical feats usually fail to call up any memories of the places themselves; memory is intertwined with history, which exists only in and through words. To forget is to produce silence. The modern city did not develop from the ancient city, nor does its history have anything to do with its past. The city of today was born with industrialization and railways at a time when faith in science and technology was absolute. It was bloated with the belief that it knew everything and was packed with people who had to be locked off the streets and placed in factories and schools and kept amused in nightclubs and bordellos.

The modern city has changed scales. It has grown into "the big city" that Georg Simmel studied and helped us to discover by following in the footsteps of this new character, the individual (Simmel, 1989). In urban space, the new individual is able to find whatever is needed to quench his or her thirst for novelty.

Entering the City

The individual stands in the city in comfortable anonymity, free to dare or to beware, spend or spare, stay put or experience all that is possible. The big city is a refuge, a hiding place, a resting place on the long road of life traveled by the being-for-death, at his or her own pace, on the individual's own scale. The individual has just arrived in the city, has just left his or her native village. The individual is afraid and hardly knows how to behave, how to act, how to speak. The individual knows nothing; for him or her, the city is a foreign land, and he or she is a stranger in these parts and feels it. The individual sees the surprised, sometimes even hostile glances of people sizing up his or her clothes and appearance. The individual says nothing and dares not move except among fellow countrymen, former neighbors, and village people with the same background, the same values, the same way of talking about life and everyday concerns, the same way of expressing permanence and absence as well as death and continuity, the same way of grieving for parents and marrying off the children. So, the individual is there, after all. A little lost in the throng of activity, but there, and bearing the weight of fate, a burden that makes the individual lose his or her footing yet also gives the individual freedom.

Guided by chance, the individual in the city wanders about as if in a drunken stupor, like a dark cloud in a stormy sky that bespeaks difficult choices. The urban individual must come to terms with his or her own life, and the individual is alone to do so, something that both delights and distresses him or her. The individual can do and decide anything, everything. He or she floats on uncertain waters but has not yet learned to swim. The individual looks for a safe haven, a buoy, a lighthouse, a pier, something to hold on to, somewhere to settle and prosper. The individual will either sink or swim, lose the battle or win both his or her world and the whole world. The urban individual is alone despite the fact of being surrounded by and immersed in love. Does she know it?

"When I arrived in Bombay, I was [18 years old]. I went to see an uncle of mine who lived in a hut in a shantytown, and I hated that city. Even though it was the first time I'd seen the sea. . . . Not only the sea, but such an incredible number of automobiles, and everywhere things to buy, and young and beautiful women . . . yes, so many extras." The young social worker in a village just outside of Pune hardly suspects that those "extras" are what in fact most interested me. They give forcefulness and intensity to what he is saying.

Let us briefly sum up the story. In the village that the individual came from, all was crystal clear. The rich were rich, and the poor were poor. The children of the poor hoped to make their way in the world, and so the individual left, only to experience a great disappointment; the big city, so full of promises, is just a humbug. There is something fake even in beauty. The sea itself does not look like what the individual imagined it to be. The city is there—in short, a display of modernity. One is free there, yes, even to drown.

In their remarkable work, Guy Poitevin and Heima Rairkar collected the stories of women who came to the city and were able, despite everything, to adapt (Poitevin & Rairkar, 1994).[3] The phrase *despite everything* is very significant. It sums up the hundreds of difficulties these women had to deal with every day—sexual harassment at work, daily survival, their neighbors' envy, the slow process of awareness, the winding path leading to autonomy. Who knows how many sink before ever seeing land? How many anonymous failures are hidden in the city's folds? How many battles, although lost in advance, are nonetheless adamantly and bravely fought? The city is ruthless. No one knows how many have been left by the wayside.

The city, or more accurately the urban territory, is an ever stretching piece of land, constantly modified by migrations, investments, and land speculation, constantly adjusting to as many desires as expectations. It does so for all reasons and for no reason at all. The city has devoured and wasted so much talent that it would be indecent not to acknowledge it. The city is our present. It is both a moment of time and a gift. It is *the* present, the root of the future, and it also is *a* present, something given to us. Who can deny ever having received such a present, that glance or friendly gesture offered by one person to another as they pass each other on the street?

Urban Lifestyles

If, as Françoise Choay writes, the "urban" has killed the city; if the city has failed to fend off the repeated assaults of technology and demographic growth and has dissolved within a larger area whose limits are vague and fluid, at once close and distant; if the city, soul and body, has disintegrated into the "urban," then what are we to do about (and inside) this "urban entity" (Choay, 1994a)?[4] It is supposed to be the result of urbanization, yet urbanization is more than a demographic, quantitative phenomenon; it is civilization, and civilization determines our ways of living.

In this sense, urbanization also has an impact on rural populations, although in a more indirect fashion. Urban models are broadcast in many ways. Migrants going back to their villages for religious feasts or family reunions do not return home empty-handed and empty-headed. Not only do they bring new tools (electrical appliances and instruments of all sorts), they also bring home new ways of doing things (e.g., cooking, cutting clothes, organizing festivities, doing handywork about the house) and new social norms (e.g., relationships with neighbors, involvement in local events, participation in associations). These new ways are revered because they are "city ways." They are accepted because they are urban, in other words, modern and civilized. Villagers often are self-deprecating, consider themselves inferior, and so readily adopt "new ways."

Eric de Rosny, studying the new beliefs arising from the excessively rapid urbanization process, describes how some people found refuge in visiting healers or priests who helped them deal with adversity. These cults represented the *idea* that city dwellers had of tradition, which in itself made them traditional, whereas in fact it was the rural dwellers who were adopting them because they were urban and "modern" (de Rosny, 1996).

Television, with its daily production of stereotypes, ready-made phrases, and attitudes, also is responsible for the *urbanization of lifestyles.* We all are aware of the impact of popular television series on parents' choices of names for their children, but there is more to it than that. Brazilian television series are de facto in charge of family planning propaganda; today, on average, Brazilian women "give birth to half as many children as they did 25 years ago (2.52 in 1995 against 5.76 in 1970)" (Sevilla, 1996). Similarly, television commercials, shown over and over, promote products that perhaps no one has yet used in the village, but in so doing, they reveal bodies, gestures, clothes, behavior, a whole world that impresses itself on the viewers.

A comparative analysis of the role played by television in the promotion of urban lifestyles certainly would be most useful. Several studies already have confirmed the insidious and decisive way in which television influences the mind. To give an example, consider a commercial on television in Cameroon. A young man in Western dress serves a beer to a young woman wearing makeup, a skirt, and a blouse. Both are smiling. The man sets the bottle down on a low table, next to a bowl of salted peanuts, and then puts his arm around her. The music is of the international type, that which cannot be identified as belonging to any specific culture. The two young people are not wearing traditional dress, are not drinking a local beverage, and do not follow the usual rules concerning touching. They live in a world that will become accessible to the viewers if they buy this brand of beer. More than just encouraging us to have a drink, the commercial is telling us how wonderful it is to live in the city.

Here is another commercial, seen in India. A national car manufacturer presenting its most recent model shows a middle class family—a father, a mother, and their two children. The boy and girl, about 10 or 12 years old, emerge from the car and run out onto the beach, laughing. The ideal size of the typical family is based on an apt balance between both genders. When seen in a village where no one owns a car, such a commercial has numerous implicit meanings that even the person who produced it would be incapable of interpreting.

Indeed, urban civilization is apprehended by the eye first. American sociologist Richard Sennett underscores the role of sight, of the eye that embraces the urban landscape as it takes in situations, educates through imitation, and assesses a distance or a relationship (Sennett, 1992, 1994). Newcomers learn to know the city with their eyes; furthermore, it is the images that they have recorded in their minds for years that help them to decode what they now discover. The city is, above all, something that is seen, which demonstrates how strong the impact of television and images can be. Only afterward does the city speak for itself and cause others to speak about it.[5] City dwellers even create a new vision of rural life that somehow seems true, picturesque, and authentic. Authenticity itself is a value created by urban culture.

Tourism also exports the big city spirit far beyond its borders, a subject that would require an entire guidebook. Thus, lifestyles are increasingly urbanized, although more or less so—more or less fast and more or less complete, depending on where we happen to find ourselves on this rapidly urbanizing planet. This *more or less* is perhaps what we should now explain a little more. First, note that this fast-growing urban civilization follows the pattern of merchant capitalism and embraces the same criteria of modernity (or "Westernization" to some). Thus, urban civilization feeds on the globalization of an economic system based on individual freedom and market forces and flourishes due to the dynamics of modernity—flows, not stocks; imbalance, not repetition; novelty, not durability; uncertainness, not stability.

Eating, Housing, Believing

Let us look at a few cases in which the urban civilization and the accompanying globalization phenomena influence lifestyles—food, home, gods, sexuality, family relationships, ambition, savings. All of these human activities differ considerably depending on where one is. In the country, the way in which we eat depends on regional specificities, the type of land ownership, the organization of farm labor, and on a host of other elements that identify a rural community. One could draw a map of regional forms of eating behavior, adding other factors (e.g., size of the family, type of instruments used, tenure type). The result would be an anthropological map of eating habits. There is a France where people eat butter and another France where olive oil is used instead.

With urbanization, eating, shopping, and cooking practices are the same for all; all city dwellers eat three times a day (even though the chosen times of day might differ from one social category to the next or from one city to the next), buy a large variety of fruits and vegetables at the market in any season, and are unable to devote much time to cooking and so tend to prepare simple dishes or eat ready-made food (canned or frozen). City cooking is a mix between country cooking, with the ingredients available in the city, and foreign dishes that have quietly become part of local history. Due to the development of the chicken industry, for example, chicken has replaced meat in many typical dishes. Wheat bread has, to a large extent, replaced the more rustic rice cakes or other grain breads produced by local food industries.[6] Working people cannot always eat at home or bring their own lunches to work. As a result, they often eat out in canteens, cafeterias, restaurants, and cafes. Consequently, the types and portions of food become more and more standard. Eating at home requires dishes and furniture, and it entails organization of labor; all of these differ from their rural and traditional counterparts.

The act of cooking goes along with an "art of the dinner table" as well as an art of spending time together. A meal prepared "like back home on the ranch" is a feast. Its role is to strengthen one's ties to a region of origin, a community, a feeling that living in the city has blunted. The symbolism is very powerful here.

The same can be said about housing conditions. The way in which people organize their homes is far from neutral; layout, furniture, and decorations all have both a practical and a ritual purpose. An apartment in a multifamily building is very different from a single-family home or a farmhouse. Even if an apartment is considered more convenient, its occupants must live in it according to built-in rules. Each of the rooms in an apartment of the "international" type has a separate function, and one must not switch or mix them. No space has been provided for the gods who protect the household, no system exists to let out the smoke from the brazier, and so on.

The size of the rooms and the size of the apartment are designed to house a specific number of inhabitants (although overcrowding is quite frequent) and to determine the way in which space is meant to be used (although private space often is appropriated in many different ways at once). For the vast majority of city dwellers (i.e., those who live in shantytowns), poor housing, lack of public services and utilities, and general insecurity are the context in which they must adapt to urban living, a context in which elements of the rural world coexist with a few pale reminders of the modern lifestyle to which they aspire.

The urbanization of housing habits, like that of eating habits, is not a single-track process, and we should not allow the description of overall tendencies to hide the diversity of actual practices.[7] Cultures are constantly interacting, blending, and intertwining. Syncretism thrives, and this brings us to our third example—the gods.

Liturgy, religious rite, and worship have their own time scale that does not correspond to that of city life. Sacred time is increasingly in conflict with secular time, resulting in many situations of malaise.[8] Synthesis can easily lead to oversimplification. In the introduction to his essay, *Le Sacré et le profane* (*The Sacred and the Profane*), Mircea Eliade writes, "The profane world as a totality, the totally desacralized Cosmos is a recent discovery of the human mind" (Eliade, 1965, p. 19). One could add that the disillusion produced by the urban environment creates new and diverse situations according to the specific characteristics of each urbanization process. Generally speaking, we observe a dual pattern; beliefs are becoming more intimate and individualized, and forms of worship are being increasingly modernized. In some cases, city life makes it impossible to hold traditional funerals that last several days and nights, involve priests and musicians, and bring together the whole extended family; in others, religious feasts occurring on Thursdays are celebrated on Sundays so that people will not miss work.

"Settling in a territory means consecrating it," writes Eliade (1965, p. 36). True enough; new residents of Seoul sometimes call in a shaman to honor the god that protects the new house, but most city dwellers have lost this sense of intimacy with the gods, a relationship based on awe and a quest for meaning. Many live in a private space devoid of meaning, with no hierophany at all, to quote Eliade once again, who defines the term as "something sacred appears before us" (p. 17).

Isaac Joseph observes, "The break with rural origins leads to a feeling of loss and a sense of fragmentation: of loss, because the world ceases to be a familiar place, and of fragmentation, because the space one lives in and the time one lives through are fragmented" (Joseph, 1984). We might add that secular and urban time scales impose themselves on us because they are governed by the principle of economy, a principle that no longer is confined to the domestic sphere or controlled by religion. Economy dominates the city and the world. *Homo urbanus* is first a *homo economicus.* Of course, we must not hastily infer from this that cities no longer have room for belief, for the sacred and the divine; they are simply not expressed in the same way as they were in rural societies or in the cities of Ancient Greece. The rationalism of urbanists and the functionalism of architects, all so full of good intentions and convinced that they are creating worlds of happiness when they build apartment blocks and housing projects, have little control over a pedestrian's imagination, an adolescent's dreams, or the magic of unexpected encounters that the city's internal dynamics are wont to provoke.

To Be or Not to Be: Inhabiting the City

As a place where the individual is able to come into his or her own, the city also creates a feeling of malaise.[9] Finding one's bearings in the modern city's intricate networks and various architectures, real or imaginary, is no easy matter. A city dweller can feel ill at ease in the city's fragmented time and space, just as he or she can feel bereft of a sense of community. Little does it matter that it is thanks to the city that the individual is free of the burden of inherited rural traditions and in a position to weave the cozy web of social relationships whereby the city dweller can play out the different aspects of his or her personality. Making that happen, of course, is not his or her responsibility alone; it also is the responsibility of urban policy in a democratic framework, and that is where problems arise. *Urban* does not necessarily mean *democratic,* and looking at an atlas of world cities tells us as much.

Some urban implosions have even been the pretext to put a stop to any attempt at democratization. Quiet please! Public opinion is being muzzled here, straitjacketed there, turned into a two-faced bully—censorship on one side and self-censorship on the other. A solitary slogan wails on a wall. Who can hear its cries? Who can tell why it weeps? In the Internet era, freedom is a novel idea. Never forced on people, it overcomes them like an overflowing shower of joy. Contagious and stubborn, even when wounded, it never gives in. It befriends the excluded, fights injustice, learns doomed languages, and inspires artists. It outsmarts prohibition and bursts into sunny song. What do we know about where these dynamics might lead?

The reality of urbanization no longer corresponds to the old scheme of the same cause having the same effects; urban diversity is fostered by a lack of discipline. This is "proof" that we are reaching, by many cultural paths, a new civilization that is transcending the oppositions between socialism and capitalism, between postindustrial society and consumer society, between developing countries and developed countries. This new civilization is one of dispersed "fragments" reunited in cities.

To inhabit, then, appears to be a decisive and vital act. We can hear Martin Heidegger whispering, "To inhabit means to be." Inhabiting the world, *our* world (that which surrounds us and that which haunts us), becomes an existential necessity for each and every person on this planet. Indeed, the very ability to speak of each person is due to the urbanization of lifestyles, itself the consequence of cultural blending, technological revolutions, the globalization of the economy and of telecommunication networks, the relations between states and nations, the relations between regional and local authorities. *Inhabiting* never can be simple; the very meaning of the verb is rooted in complexity. *Inhabiting* is another name for living with, doing together, thinking among. *Inhabiting* is not something one learns. *Inhabiting* is a risk and a risk we must run by refusing to live in a cellophane-wrapped world, in prepackaged housing with partners selected by a computer program, by refusing to be tested for jobs, and so on. *Inhabiting* means, first and foremost, controlling one's own possibilities to be precise—as the Thom(p)sons might say, one's own impossibilities as well.[10] *To inhabit* means to dwell in the wavering truth of one's desires, in the unlikely company of one's imaginary ancestors, and to grow cultural and linguistic roots and join the network. Urban freedom breaks loose and befriends the wind.

"Come to me, and I will come to you," cries *homo urbanus.* The urban city is an encounter over which determinism has no hold. It is a place of friendship. The "urban" wants to be amicable, and the architecture wishes to be likable, precisely to foster friendship. *Friendship* is another word for *knowledge;* it is situated somewhere between Eros and Agapè. Somewhere between the desire to conquer and the discovery of a priceless, dual form of humility—respect for others and acceptance of one's own strangeness. To paraphrase something Heidegger once said, to take possession is to become "other" through contact with others. What a wonderful prospect.

With the urbanization of the planet, the world as we know it is radically changing, although we hardly know how or into what. Relationships between city and countryside are entering a new phase, during which the country too will become urbanized and the city will undergo deep transformations. Urban lifestyles and the values they represent are spreading to society as a whole and are becoming increasingly international. Consequently, the urban challenge arouses both fears and hopes. So far, there is no evidence that the urban future will necessarily look like *1984*. On the contrary, the world might very well become a more human place, one where the individuality of the being-in-the-world can assert itself more fully and where cultures learn to communicate. Because imagining the worst is something everyone can do, let us for a change think plural, open, possible, infinite.

NOTES

1. See also, in particular, my 1996 paper, "Lieu, hors-lieu, et être-au-monde."

2. In 1670, Furetière defined the city in his dictionary as follows: "a rather large human settlement, usually surrounded by walls." (This definition is from a famous dictionary, *Dictionnaire,* written by Antoine Furetière, a French novelist, satirist, and lexicographer.)

3. This remarkable survey studies how village women arrive in the city, their relationship to paid jobs, and how they settle in shantytowns, among other topics. It is an extremely interesting and informative piece of work. See also the preface by Thierry Paquot.

4. See also Choay (1994b, 1994c).

5. An entire area of research is devoted to the microsociological study of interpersonal relationships in the urban environment. See Chombart de Lauwe (1996), Goffman (1956), Gutwirth and Petonnet (1987), Joseph (1984), Moles and Rohmer (1978), Petonnet (1993), Remy (1995), and Simmel (1989).

6. Concerning the relationships between geography and cultures, see the excellent book by Gourou (1984). Concerning urbanization and eating habits, see Ariès (1997), Desjardins (1985), O'Deye (1985), and Pynson (1987).

7. See "Anthropologie de l'espace habité" (1992), Ascher (1995), Culturello (1992), Deffontaines (1972), Haumont and Segaud (1989), Matras-Guin and Taillard (1992), Poitevin and Rairkar (1994), and Zouilai (1990) on housing and Islam. See also the special issue of *Urbanisme* on "Habitat" (1998, No. 298).

8. See Eliade (1965), Costa-Lascoux and Yu-Sion (1995), Paquot (1990), Poitevin and Rairkar (1994), Racine (1993), and Robuchon (1993). See also "Les villes, lieux privilegiés de la mission" in *Spiritus* (September 1994, No. 136) and "Chrétiens dans la ville" in *Christus* (April 1995, No. 166).

9. A considerable literature has been published on this subject by Hannah Arendt, Emmanuel Levinas, Martin Heidegger, and Henri Lefèbvre, among other important

writers. Some of these texts are analyzed in my *Vive la ville!* (Paquot, 1994). One also might read Breton (1995), Maldiney (1975), Taylor (1992), and Younès and Mangematin (1996).

10. Thomson and Thompson are the comic police detectives in the "Adventures of Tintin" comic book series by Hergé. They are constantly muddled, and one always is repeating the words of the other, adding "to be precise."

REFERENCES

Anthropologie de l'espace habité. (1992). *L'Homme et la Société*, No. 104.

Ariès, P. (1997). *Les fils de McDO: La McDonaldisation du monde.* Paris: L'Harmattan.

Ascher, F. (Ed.). (1995). *Le logement en question.* La Tour d'Aigues, France: De l'Aube.

Auge, M. (1992). *Non-lieux: Introduction à une anthropologie de la surmodernité.* Paris: Seuil.

Braudel, F. (1967). *Civilisation matérielle et capitalisme.* Paris: Armand Colin.

Breton, S. (1995). *L'autre et l'ailleurs.* Paris: Descartes & Cie.

Choay, F. (1994a). Le règne de l'urbain et la mort de la ville. In J. Dethier & A. Guiheux (Eds.), *La Ville: Art et architecture en Europe, 1870-1993.* Paris: Éditions du Centre Pompidou.

Choay, F. (1994b). Penser la non-ville et la non-campagne de demain. In *La France au-delà du siècle* (pp. 23-32). La Tour d'Aigues, France: De l'Aube.

Choay, F. (1994c). Six thèses en guise de contribution: Une réflexion sur les échelles d'aménagement et le destin des villes. In A. Berque (Ed.), *La maîtrise de la ville, urbanité française, urbanité nippone.* Paris: École des Hautes Études en Sciences Sociales.

Chombart de Lauwe, P.-H. (1996). *Un anthropologue dans le siècle.* Paris: Descartes & Cie.

Costa-Lascoux, J., & Yu-Sion, L. (1995). *Paris XII, lumières d'Asie.* Paris: Autrement.

Coupland, D. (1996). *Microserfs.* New York: HarperCollins.

Culturello, P. (1992). *Regards sur le logement.* Paris: L'Harmattan.

Deffontaines, P. (1972). *L'Homme et sa maison.* Paris: Gallimard.

de Rosny, E. (1996). La persistance des rites dans l'Afrique urbaine. *Urbanisme,* No. 291.

Desjardins, D. R. (1985). Urbanisation et évolution des modèles alimentaires: L'exemple de la Côte d'Ivoire. In T. Paquot (Ed.), *Nourir les villes en Afrique sub-Saharienne.* Paris: L'Harmattan.

Echenoz, J. (1997). *Un an.* Paris: Les Éditions de Minuit.

Eliade, M. (1965). *Le sacré et le profane.* Paris: Gallimard.

Goffman, E. (1956). *The presentation of self in everyday life* (2 vols.). New York: Doubleday.

Gourou, P. (1984). *Riz et civilisation.* Paris: Fayard.

Gutwirth, J., & Petonnet, C. (Eds.). (1987). *Chemins de la ville, enquêtes ethnologiques.* Paris: Éditions du CTHS.

Haumont, N., & Segaud, M. (Eds.). (1989). *Familles, modes de vie et habitat.* Paris: L'Harmattan.

Joseph, I. (1984). Urban anthropology also studies urban lifestyles. In *Le Passant considérable: Essai sur la dispersion de l'espace public.* Paris: Librairie des Méridiens.
Maldiney, H. (1975). *Antres de la langue et demeures de la pensée.* Lausanne, France: L'Age d'Homme.
Matras-Guin, J., & Taillard, C. (Eds.). (1992). *Habitations et habitat d'Asie du sud-est continentale.* Paris: L'Harmattan.
Moles, A., & Rohmer, E. (1978). *Psychologie de l'espace.* Tournai, France: Casterman.
O'Deye, M. (1985). A propos de l'évolution des styles alimentaires à Dakar. In *Nourrir les villes en Afrique sub-Saharienne.* Paris: L'Harmattan.
Paquot, T. (1990). *Homo urbanus.* Paris: Le Félin.
Paquot, T. (1994). *Vive la ville!* Paris: Arléa-Corlet.
Paquot, T. (1996). Lieu, hors-lieu, et être-au-monde. In C. Younès & M. Mangematin (Eds.), *Lieux contemporaines.* Paris: Descartes & Cie.
Petonnet, C. (Ed.). (1993). *Ferveurs contemporaines.* Paris: L'Harmattan.
Poitevin, G., & Rairkar, H. (1994). *Femmes coolies en Inde.* Paris: Syros.
Pynson, P. (1987). *La France à table.* Paris: La Découverte.
Racine, J.-B. (1993). *La ville entre Dieu et les hommes.* Geneva: Presses Bibliques Universitaires.
Remy, J. (Ed.). (1995). *G. Simmel: Ville et modernité.* Paris: L'Harmattan.
Robuchon, G. (1993). Etagères à bons dieux: Autels domestiques tamouls en immigration. In *Ferveurs contemporaines.* Paris: L'Harmattan.
Sennett, R. (1992). *The conscience of the eye: The design and social life of cities.* New York: Norton.
Sennett, R. (1994). *Flesh and stone: The body and the city in Western civilization.* New York: Norton.
Sevilla, J. (1996, October 25). La télévision brésilienne fait baisser la natalité. *Le Monde.*
Simmel, G. (1989). *Philosophie de la modernité.* Paris: Payot.
Taylor, C. (1992). *Malaise of modernity.* Cambridge, MA: Harvard University Press.
Virilio, P. (1995). *La vitesse de libération.* Paris: Galilée.
Virilio, P. (1996). *Cybermonde ou la politique du pire.* Paris: Textuel.
Younès, C., & Mangematin, M. (Eds.). (1996). *Le philosophe chez l'architecte.* Paris: Descartes & Cie.
Younès, C., & Mangematin, M. (Eds.). (1997). *Lieux contemporains.* Paris: Descartes & Cie.
Zouilai, K. (1990). *Des voiles et des serrures.* Paris: L'Harmattan.

Part II

Global Perspectives

5

Whose City Is It? Globalization and the Formation of New Claims

SASKIA SASSEN

The centrality of place in a context of global processes makes possible transnational economic and political openings for the formation of new claims and, hence, for the constitution of entitlements, notably rights to place. At the limit, this could be an opening for new forms of citizenship. Within the city, these new claims are being shaped by global capital that uses the city as an organizational commodity, but they also are being shaped by disadvantaged sectors of the urban population that frequently are as internationalized a presence in large cities as is capital. The denationalizing of urban space and the formation of new claims by transnational actors raise the question, "Whose city is it?"

My argument is organized around the notion that place is central to the multiple circuits through which economic globalization is constituted. One strategic type of place for these developments, and the one focused on here, is the city. Including cities in the analysis of economic globalization, however, is not without conceptual consequences. Economic globalization has been conceptualized mostly in terms of the national-global duality such that the latter gains at the expense of the former and in terms of the internationalization of capital (and then only the upper circuits of capital).

AUTHOR'S NOTE: This chapter is a substantially revised version of an article originally published in 1996 in *Public Culture,* Volume 8, pages 205-233. It appears here with the permission of Duke University Press.

Introducing cities in an analysis of economic globalization allows us to reconceptualize processes of economic globalization as concrete economic complexes situated in specific places. Moreover, a focus on cities decomposes the nation-state into a variety of subnational components, some profoundly articulated with the global economy and others not. It also signals the declining significance of the national economy as a unitary category in a global economy. Even if, to a large extent, this was a unitary category except in political discourse and policy, it has become even less so in the past 15 years.

Why is it important that we recover place in analyses of the global economy, particularly place as constituted in major cities? The answer is because it allows us to see the multiplicity of economies and work cultures in which the global information economy is embedded. A focus on cities also captures the concrete, localized processes through which globalization exists and indicates how much of the multiculturalism in large cities is a part of globalization along with international finance. Finally, focusing on cities enables us to specify a geography of strategic places at the global scale, places bound to each other by the dynamics of economic globalization.

This is the new geography of centrality. One of the questions it engenders is whether a transnational geography also is the space for a transnational politics. Insofar as my economic analysis of the global city recovers the broad array of jobs and work cultures that are part of the global economy (although typically not marked as such), it allows me to examine the possibility of a new politics of traditionally disadvantaged actors operating in this new transnational economic geography. This politics arises out of actual participation of workers in the global economy—whether factory workers in export processing zones or cleaners on Wall Street—under conditions of disadvantage and without recognition of their participation by either corporate and media elites or the general public.

These political openings consolidate capacities across national boundaries and sharpen conflicts within such boundaries. Global capital and the new immigrant workforce are two major instances of transnationalized actors, each with unifying properties that cross borders and each in contestation with the other inside global cities. Thus, global cities are the sites for the overvalorization of corporate capital and the devalorization of disadvantaged workers. Many of the disadvantaged workers in global cities, moreover, are women, immigrants, and people of color, that is, men and women whose sense of membership is not necessarily adequately captured in national terms and, indeed, often evinces cross-border solidarities around issues of substance. My argument here is that both types of actors find in the global city a strategic site for their economic and political activities.

The analysis presented here grounds its interpretation of the new politics induced by globalization in a detailed understanding of the economics of globalization and, specifically, in the centrality of place, that is, in a context where place typically is seen as neutralized by the capacity for global communications and control. My first assumption is that it is important to dissect the economics of globalization to understand whether a new transnational politics can be centered in the new transnational economic geography. Second, dissecting the economics of place in the global economy allows us to recover the noncorporate components of economic globalization and to inquire about the possibility of a new type of transnational politics. Is there a transnational politics embedded in the centrality of place and in the new geography of strategic places, that is, in the new worldwide grid of global cities? Immigration, for example, is one such process through which a new transnational political economy is being constituted. It also is largely embedded in major cities because most immigrants—whether in the United States, Japan, or Western Europe—are concentrated there. As one of the constitutive processes of globalization today, immigration cannot be ignored in any account of the global economy.

■ Place and Work Process in the Global Economy

The mainstream account of economic globalization is a narrative of eviction. Key concepts in that account—globalization, information economy, and telematics—all suggest that place no longer matters and that the only worker who matters is the highly educated professional. This account privileges global communications over the material infrastructure that makes such communication possible; it privileges information outputs over the workers producing those outputs, from specialists to secretaries. It also privileges the new transnational corporate culture over the multiplicity of work cultures, including immigrant cultures, within which many of the "other" jobs of the global information economy take place. In brief, the dominant narrative concerns itself with the upper circuits of capital, particularly the hypermobility of capital, rather than with that which is place bound.

Massive trends toward the spatial dispersal of economic activities at the metropolitan, national, and global levels are indeed taking place, but they represent only half of what is happening. Alongside the well-documented spatial dispersal of economic activities, new forms of territorial centralization of top-level management and control operations have appeared. National and global markets as well as globally integrated operations require central places where the work of globalization is done (Friedmann, 1995; Sassen, 1991). Furthermore, information industries require a vast physical infrastructure containing strategic nodes with hyperconcentrations of facilities (Castells, 1989; Graham & Marvin, 1995). Finally, even the most advanced information industries have a work process, that is, a complex of workers, machines, and buildings that are more place bound and more diversified in their labor inputs than the imagery of information outputs suggests (for general discussions of this approach, see Friedmann, 1995; Sachar, 1990; and Stren, 1996).

Centralized control and management over a geographically dispersed array of economic operations does not come about automatically as part of a "world system." It requires the production of a vast range of highly specialized services, telecommunications infrastructure, and industrial services that are crucial for the valorization of today's leading components of capital. A focus on place and work process turns our attention away from the power of large corporations over governments and economies to the range of activities and organizational arrangements necessary for the implementation and maintenance of a global network of factories, service operations, and markets. These processes are only partly encompassed by the activities of transnational corporations and banks.

One of the central themes of my work has been cities as production sites for the leading service industries of our time. My intent has been to identify the infrastructure of activities, firms, and jobs that are necessary to run the advanced corporate economy. The focus is on the *practice* of global control—the work of producing and reproducing, under conditions of economic concentration, the organization and management of a global production system and a global marketplace for finance (Sassen, 1991, 1994). This, in turn, allows a focus on the infrastructure of jobs involved in this production including low-wage, unskilled manual jobs typically not thought of as being part of advanced globalized sectors.

Global cities are centers for the *servicing* and *financing* of international trade, investment, and headquarters operations; that is, the multiplicity of specialized activities present in global cities are now crucial in the valorization, indeed the overvalorization, of leading sectors of capital. In this sense, they are strategic production sites for today's leading economic sectors. This function is reflected in the ascendance of these activities in their economies, albeit with significant local variations (Abu-Lughod, 1995; *Le Débat,* 1994; Noller, Prigge, & Ronneberger, 1994; von Petz & Schmals, 1992). Elsewhere, I have posited that what is specific about the shift to services is not merely the growth in service jobs but, most important, the growing service intensity in the organization of advanced economies (Sassen, 1994, chap. 4). Firms in all industries, from mining to wholesale, buy more accounting, legal, advertising, financial, and economic forecasting services today than they did 20 years ago. Whether at the global or regional level, cities often are the best production sites for such specialized services. The rapid growth and disproportionate concentration of such services in cities signals that the latter have reemerged as significant production sites after losing this role during the period when mass manufacturing was the dominant sector of the economy. Under mass manufacturing and Fordism, the strategic spaces of the economy were the large-scale integrated factory and the government through its Fordist/Keynesian functions.

Furthermore, the vast new economic topography that is being implemented through electronic space is one moment, one fragment, of a more vast economic chain that is, in good part, embedded in nonelectronic spaces. There is no such thing as a fully dematerialized firm or industry. Even the most advanced information industries, such as finance, are installed only partly in electronic space, as are industries that produce digital products such as software. The growing digitalization of economic activities has not eliminated the need for major international business and financial centers and all the material resources they concentrate, from state-of-the-art telematics infrastructure to brain talent (Castells, 1989; Graham & Marvin, 1995; Sassen, 1994).[1]

It is precisely because of the territorial dispersal facilitated by telecommunications advances that agglomeration of centralizing activities has expanded immensely. This is not a mere continuation of old patterns of agglomeration but rather a new agglomeration logic. Many of the leading sectors of the economy operate globally in uncertain markets, under conditions of rapid change in other countries (e.g., deregulation, privatization), and subject to enormous speculative pressures. What glues these conditions together in a new logic for spatial agglomeration is the added pressure of speed.

A focus on the *work* behind the command functions, on the actual *production process* in the finance and services complex, and on global market*places* incorporates the material facilities underlying globalization and the whole infrastructure of jobs typically not marked as belonging to the corporate sector of the economy. An economic configuration very different from that suggested by the concept of "information economy" emerges. We recover the material conditions, production sites, and place-boundedness that are part of globalization and the information economy.

This approach also captures the broad range of firms, workers, work cultures, and residential milieus that comprise globalization processes although never marked, recognized, represented, or even valorized as such. In this regard, the new urban economy is highly problematic. This is particularly evident in global cities and their regional counterparts, where a whole series of new dynamics of inequality is set in motion (Fainstein, Gordon, & Harloe, 1993; Peraldi & Perrin, 1996; Sassen, 1994, chap. 5). The new growth sectors, specialized services and finance, have profit-making capacities vastly superior to those of more traditional economic sectors. Although many of the latter remain essential to the operation of the urban economy and the daily needs of residents, their survival is threatened when finance and specialized services earn super-profits and bid up prices.[2] Polarization in the profit-making capabilities of different sectors of the economy always has existed. What is happening today, however, takes place at another order of magnitude and is causing massive distortions in the operations of various markets, from housing to labor (Fainstein et al., 1993; Hitz et al., 1995). We can see this effect, for example, in the retreat of many real estate developers from the low- and medium-income housing market in the wake of the rapidly expanding housing demand by the new highly paid professionals and the possibility for vast overpricing of this housing supply.

A dynamic of valorization has sharply increased the distance between the valorized, indeed the overvalorized, sectors of the economy and the devalorized sectors, even when the latter are part of leading global industries. This devalorization of growing economic sectors is embedded in a massive demographic transition toward a growing presence of women, African Americans, and Third World immigrants in the urban workforce, a subject I return to later.

We see here an interesting correspondence between great concentrations of corporate power and large concentrations of "others." Large cities in the highly developed world are the terrain in which a multiplicity of globalization processes assume concrete and localized forms.

By focusing on cities, we can capture not only the upper but also the lower circuits of this globalization. Localized forms are, in good part, what globalization is about. We can then think of cities as one of the sites for the contradictions of the internationalization of capital. Furthermore, if we consider that large cities also concentrate a growing share of disadvantaged populations—immigrants in Europe and the United States, African Americans and Latinos in the United States—then it becomes obvious that cities have become a strategic terrain for a whole series of conflicts and contradictions.

■ A New Geography of Centrality and Marginality

The global economy materializes in a worldwide grid of strategic places, uppermost among which are major international business and financial centers. This global grid constitutes a new economic geography of centrality, one that cuts across national boundaries and across the old north-south divide. It has emerged as a parallel political geography, a transnational space for the formation of new claims by global capital; I return to this in the next section.

This new economic geography of centrality partly reproduces existing inequalities but also is the outcome of a dynamic specific to the current forms of economic growth. It assumes many forms and operates in many terrains, from the distribution of telecommunications facilities to the structure of the economy and of employment. Global cities are sites for immense concentrations of economic power and exist as command centers in a global economy, whereas cities that once were major manufacturing centers (e.g., Detroit, Manchester) have suffered inordinate declines.

The most powerful of these new geographies of centrality at the interurban level binds the major international financial and business centers—New York, London, Tokyo, Paris, Frankfurt, Zurich, Amsterdam, Los Angeles, Sydney, and Hong Kong, among others. Now, this geography also includes cities such as São Paulo, Buenos Aires, Bombay, Bangkok, Taipei, and Mexico City. The intensity of transactions among these cities—particularly through financial markets, transactions in services, and investments—has increased sharply, and so have the orders of magnitude involved. At the same time, there has been a sharpening inequality in the concentration of strategic resources and activities between each of these cities and others in the same country.

One might have expected that the growing number of financial centers now integrated into the global markets would have reduced the extent of concentration of financial activity in the top centers, particularly given the immense increases in the global volume of transactions. Yet, the levels of concentration remain unchanged in the face of massive transformations in the financial industry and in the technological infrastructure on which this industry depends.[3]

Furthermore, this unchanged level of concentration has happened at a time when financial services are more mobile than ever before. In the context of massive advances in telecommunications and electronic networks, globalization, deregulation (an essential ingredient for globalization), and securitization have been the key to this mobility. One result is growing competition among centers for hypermobile financial activity. In my view, competition has been overemphasized in specialized accounts on this subject. As I have argued elsewhere (Sassen, 1991, chap. 7), a functional division of labor exists among various major financial centers. In that sense, we can think of a transnational system with multiple locations.

The growth of global markets for finance and specialized services, the need for transnational servicing networks due to sharp increases in international investment, the reduced role of the government in the regulation of international economic activity, and the corresponding ascendance of other institutional arenas, notably global markets and corporate headquarters, all point to the existence of transnational economic processes with multiple locations in more than one country. Here is the formation, at least incipient, of a transnational urban system. Not only do these cities compete with each other, but a division of labor also exists in this cross-border network of global cities.

Alongside these new global and regional hierarchies of cities is a vast territory that has become increasingly peripheral, excluded from the major economic processes that fuel economic growth in the new global economy. A number of formerly important manufacturing centers and port cities have lost functions and are in decline, not only in the less developed countries but also in the most advanced economies. This is yet another meaning of economic globalization.

Inside global cities as well, we see a new geography of centrality and marginality. The downtowns of cities and key nodes in metropolitan areas receive massive investments in real estate and telecommunications, whereas low-income city areas and the older suburbs are starved for resources. The incomes of highly educated workers rise to unusually high levels, and those of low- or medium-skilled workers sink. Financial services produce super-profits, whereas industrial services barely survive. These trends are evident at different levels of intensity in a growing number of major cities in the developed world and more and more in some of the developing countries that have been integrated into the global financial markets (Cohen, Ruble, Tulchin, & Garland, 1996; Kowarick & Campanario, 1986; Sassen, 1994, chap. 2; but see also Simon, 1995).

The Formation of Global Rights for Capital in the New Urban Grid

The preceding analysis points to a space economy for major new transnational economic processes that diverges in significant ways from the international-national duality presupposed in many analyses. Economic globalization does indeed extend the economy beyond the boundaries of the nation-state and, hence, reduces the state's sovereignty over its economy, an observation that now is a basic proposition in discussions of the global economy. This is particularly evident in the leading information industries. Existing systems of governance and accountability for transnational economic activities and entities leave much ungoverned when it comes to these industries. Global markets in finance and advanced services operate partly through a regulatory umbrella that is not state centered but rather market centered. On the other hand, a focus on the space economy of information industries shows us that important components of these industries are embedded in particular sites within national territories. Thus, even the most digitalized and globalized industries are at least partly embedded in national institutional frameworks and geographic terrains.

A strategic subnational unit such as the global city illuminates the two conditions, global reach and place specificity, that are at opposite ends of the governance challenge posed by globalization but are not captured in the more conventional national-global duality. A focus on leading information industries in global cities introduces into the discussion of governance the possibility that the capacity for regulation can stem from the concentration of significant resources, including fixed capital, in strategic places. The considerable place-boundedness of many of these resources contrasts with the hypermobility of information outputs. The regulatory capacity of the state stands in a different relation to hypermobile outputs than to the infrastructure of facilities, from office buildings served by fiber-optic cable to specialized workforces, present in global cities.

At the other extreme, the fact that many of these industries operate in global markets that are partly electronic raises questions of control that derive from key properties of the new information technologies, notably the orders of magnitude in trading volumes made possible by speed. Here, it no longer is just a question of the capacity of the state to govern these processes but also a question of the capacity of the private sector, that is, of the major actors involved in setting up these markets in electronic space. Elementary and well-known illustrations of this issue of control are stock market crashes attributed to program trading and globally implemented decisions to invest or disinvest in a currency or an emerging market that resembles a sort of worldwide stampede facilitated by the fact of global integration and instantaneous execution worldwide.[4]

The specific issues raised by these two variables (i.e., place-boundedness and global reach) are quite distinct from those typically raised in the context of the national-global duality.[5] A focus on this duality leads to rather straightforward propositions about the declining significance of the state vis-à-vis global economic actors. The overarching tendency in economic analyses of globalization and of information industries has been to emphasize certain aspects—industry outputs rather than the work process involved, the capacity for instantaneous transmission around the world rather than the infrastructure necessary for this capacity, the impossibility of the state to regulate those outputs and that capacity insofar as they extend beyond the nation-state. By itself, this is quite correct, but it is a partial account of the implications of globalization for governance.

The transformation in the composition of the world economy, especially the rise of finance and advanced services as leading industries, is contributing to a new international economic order dominated by financial centers, global markets, and transnational firms. Correspondingly, other political categories, both sub- and supranational, have become significant.[6] Cities that function as international business and financial centers institute direct transactions with world markets without government inspection, for example, the Euro-markets or New York City's international financial zone (international banking facilities). These cities and the globally oriented markets and firms they contain mediate in the relation of the world economy to nation-states and in the relations among nation-states.

A key component in the transformation over the past 15 years has been the formation of new claims by global capital—the claim on national states to guarantee the domestic and global rights of capital. Transnational economic processes inevitably interact with systems for the governance of national economies insofar as those processes materialize in concrete places.

The hegemony of neoliberal concepts of economic relations with its strong emphasis on markets, deregulation, and free international trade has influenced policy during the 1980s in the United States, the United Kingdom, and now increasingly also in continental Europe. This has contributed to the formation of transnational legal regimes that are centered in Western economic concepts of contract and property rights.[7] Through the International Monetary Fund and the World Bank as well as the General Agreement on Trade and Tariffs (renamed the World Trade Organization as of January 1, 1995), this regime has spread to the developing world.

Deregulation and privatization are key components in the formation of these new claims by global capital. Deregulation has been a crucial mechanism to negotiate the juxtaposition of the global and the national. It is not simply about less government but also about new forms of government participation in the economy and about the participation of private entities in regulating the economy. Privatization is not simply a change in ownership regime and deregulation. It also entails a privatizing of coordination and governance functions that shift from the public to the private corporate sector, often the corporate world of major international business centers.

In brief, the strategic spaces where many global processes are embedded often are national; the mechanisms through which new legal forms, necessary for globalization, are implemented often are part of state institutions. The infrastructure that makes possible the hypermobility of financial capital at the global scale is embedded in various national territories. These developments signal a transformation in the articulation of sovereignty and territory that has marked the history of the modern state and interstate system, beginning with World War I and culminating in the Pax Americana period.

One way of describing the specific process underway today is as the "denationalizing" of national territory (Sassen, 1996, chap. 1). Both through corporate practices and through the fragments of an ascendant new legal regime, territory is being denationalized, although in specific and highly specialized ways, as befits the tenor of this era.

This denationalization cannot be reduced to a geographic conception, as was the notion in the heads of the generals who fought the wars for nationalizing territory during earlier centuries. This is a denationalizing of specific institutional arenas. Manhattan and London are the equivalents of free trade zones when it comes to finance. It is not Manhattan as a geographic entity with all its layers of activity and functions and regulations that is a free trade zone. It is a highly specialized functional or institutional realm in Manhattan that becomes denationalized. This specialized institutional realm is disproportionately concentrated in global cities.

Unmooring Identities and a New Transnational Politics

Typically, the analysis of the globalization of the economy privileges the reconstitution of capital as an internationalized presence; it emphasizes the vanguard character of this reconstitution. At the same time, it remains absolutely silent about another crucial element of this transnationalization, one that some, like myself, see as the counterpart of that of capital: the transnationalization of labor. We still are using the language of immigration to describe this process.[8] In addition, that analysis overlooks the transnationalization in the formation of identities and loyalties among various population segments that explicitly reject the imagined community of the nation. With this come new solidarities and notions of membership. Major cities have emerged as strategic sites for the transnationalization of labor and the formation of transnational identities. In this regard, they are sites for new types of political operations (Berner & Korff, 1995; Cohen et al., 1996; Dunn, 1994; *Wissenschaft Forum,* 1995).

Cities are the terrain in which a multiplicity of cultures come together and people from many different countries are most likely to meet. The international character of major cities lies not only in their telecommunications infrastructures and international firms but also in the many different cultural environments in which these workers exist. One no longer can think of centers for international business and finance simply in terms of corporate towers and corporate culture. Today's global cities are in part the spaces of postcolonialism and indeed contain conditions for the formation of a postcolonialist discourse (Hall, 1991; King, 1995).[9]

The large Western city of today concentrates diversity. Its spaces are inscribed with the dominant corporate culture but also with a multiplicity of other cultures and identities. The slippage is evident; the dominant culture can encompass only part of the city.[10] Although corporate power inscribes these cultures and identities with "otherness," thereby devaluing them, they are present everywhere. For example, through immigration, originally highly localized cultures are present in many large cities, whose elites think of themselves as cosmopolitan, that is, transcending any locality. An immense array of cultures from around the world, each rooted in a particular country or village, are reterritorialized in a few single places such as New York, Los Angeles, Paris, London, and (most recently) Tokyo.[11]

Immigration and ethnicity too often are constituted as otherness. Understanding them as a set of processes whereby global elements are localized, international labor markets are constituted, and cultures from all over the world are deterritorialized puts them at the center of the stage as a fundamental aspect of globalization (Sassen, 1998, chaps. 2-3). There has been growing recognition of the formation of an international professional class of workers and of highly internationalized environments due to the presence of foreign firms and personnel, the formation of global markets in the arts, and the international circulation of high culture. What has not been recognized is the possibility that we are seeing an internationalized labor market for low-wage manual and service workers. This process continues to be couched in terms of the "immigration story," a narrative rooted in an earlier historical period.

Certain representations of globality have not been recognized as such or are contested. Among these is the question of immigration and the multiplicity of cultural environments it creates in large cities, often subsumed under the notion of ethnicity. What we still narrate in the language of immigration and ethnicity actually is a series of processes having to do with the globalization of economic activity, cultural activity, and identity formation. Immigration and ethnicity are constituted as otherness. Understanding them as a set of processes whereby global elements are *localized,* international labor markets are constituted, and cultures from all over the world are de- and reterritorialized puts them right there at the center along with the internationalization of capital as a fundamental aspect of globalization. This way of narrating the migration events of the postwar era captures the ongoing weight of colonialism and postcolonial forms of empire on major processes of globalization and specifically on those binding emigration and immigration countries.[12] Although the specific genesis and contents of their responsibility will vary from case to case and from period to period, none of the major immigration countries is an innocent bystander.

Making Claims on the City

These processes signal a change in the linkages that bind people and places and in the corresponding formation of claims. Throughout history, people have moved and through these movements have constituted places. Today, however, the articulation of territory and people is being constituted in a radically different way in at least one regard, and that is the speed with which that articulation changes. One consequence is the expansion of the space within which actual and possible linkages can happen (Martinotti, Chapter 8, this volume). The shrinking of distance and time that characterizes the current era finds one of its most extreme forms in electronically based communities of individuals or organizations from all around the globe interacting in real time and simultaneously, as is possible through the Internet and kindred electronic networks.

Another radical form assumed by the linkage of people to territory is the loosening of identities from what have been traditional sources such as the nation or the village. This unmooring of identity formation engenders new notions of community of membership and entitlement.

The space constituted by the global grid of global cities, a space with new economic and political potentialities, is perhaps one of the most strategic spaces for the formation of transnational identities and communities. This space is place-centered; it is embedded in particular and strategic sites. It is transterritorial because it connects sites that are not geographically proximate yet are intensely connected to each other. Not only the transmigration of capital takes place in this global grid. That of people, both rich (i.e., the new transnational professional workforce) and poor (i.e., most migrant workers) do so as well. The global grid is a space for the transmigration of cultural forms and for the reterritorialization of "local" subcultures. An important question is whether it also is a space for a new politics, one going beyond the politics of culture and identity, although likely to be embedded in these.

Yet another way of thinking about the political implications of this strategic transnational space is the notion of the formation of new claims on that space. Has economic globalization at least partly shaped the formation of claims? Indeed, major new actors are making claims on these cities, notably international businesspeople and foreign firms that have been increasingly entitled to do business through progressive deregulation of national economies. These are the new city users, and they have profoundly marked the urban landscape. Their claims to the city are not contested, even though the costs and benefits to cities have barely been examined. One result is the incipient denationalization dynamics discussed in the previous section that, although institutional, tend to have spatial outcomes disproportionately concentrated in global cities.

City users have made an often immense claim on the city and have reconstituted strategic spaces of the city in their image. There is a de facto claim to the city that never is problematized. New city users change the social morphology of the city and constitute what Guido Martinotti calls the metropolis of second generation, the city of late modernism. Martinotti (1993) notes that the new city of city users is a fragile one whose survival and successes are centered on an economy of high productivity, advanced technologies, and intensified exchanges.

On the one hand, for international businesspeople, the city as a space consists of airports, top-level business districts, luxury hotels and restaurants, a sort of urban glamour zone. On the other hand, does a city that functions as an international business center in fact recover the costs involved in being such a center—the costs involved in maintaining a state-of-the-art business district and all that it requires, from advanced communications facilities to top-level security and "world-class" culture?

At the other extreme are those who use urban political violence to make their claims on the city, claims that lack the de facto legitimacy enjoyed by the new city users. These claims are made by actors struggling for recognition, entitlement, and rights to the city. Sophie Body-Gendrot shows how the city remains a terrain for contest, characterized by the emergence of new actors, often increasingly younger. It is a terrain in which the constraints placed on, and the institutional limitations of, governments to address the demands for equity engender social disorders. Body-Gendrot (1993) argues that urban political violence should be interpreted not as a coherent ideology but rather as an element of temporary political tactics that permits vulnerable actors to interact with the holders of power on terms that will be more favorable to the weak.

Two aspects in this formation of new claims have implications for the new transnational politics. One is the sharp and perhaps sharpening differences in the representation of these claims by different sectors, notably international business and the vast population of low-income others—African Americans, immigrants, women. The second aspect is the increasingly transnational element in each of these types of claims and claimants. It signals a politics of contestation embedded in specific global cities but transnational in character. At its extreme, the difference assumes the form of (a) an overvalorized corporate center occupying a smaller terrain whose edges are sharper than, for example, those during the postwar era with its large middle class and (b) a sharp devalorization of what is outside the center, which comes to be read as marginal. The difference repeats itself in particular local forms and contents from one city after another in the global grid. In this cross-border reverberation lies a de facto transnationalism, perhaps illustrated in the social movements described by Margit Mayer (Chapter 10 of this volume)—movements about similar issues but in different places.

A question here is whether the growing presence of immigrants, African Americans, and women in the labor forces of large cities is what has facilitated the embedding of this sharp increase in inequality (as expressed culturally and in earnings). The new politics of identity and the new cultural politics have brought many of these devalorized or marginal sectors into the forefront of urban life.[13]

Conclusion

Globalization is a contradictory space. It is characterized by contestation, internal differentiation, and continuous border crossings. The global city is emblematic of this condition. The global city concentrates a disproportionate share of global corporate power and is one of the key sites for its overvalorization. It also concentrates a disproportionate share of the disadvantaged and is one of the key sites for their devalorization. This joint presence happens in a context where (a) the globalization of the economy has grown sharply and cities have become increasingly strategic for global capital and (b) marginalized people have found their voice and are making claims on the city as well. This joint presence is further brought into focus by the sharpening of the distance between the two. The center now concentrates immense power, a power that rests on the capacity for global control and the ability to produce super-profits. Marginality, notwithstanding anemic economic and political power, has become an increasingly strong presence through the new politics of culture and identity. An emergent transnational politics is embedded in the new geography of economic globalization. Actors, increasingly transnational and in contestation, find in the city the strategic terrain for their operations.

NOTES

1. Telematics and globalization have emerged as fundamental forces reshaping the organization of economic space. This reshaping ranges from the spatial virtualization of a growing number of economic activities to the reconfiguration of the geography of the built environment for economic activity. Whether in electronic space or in the geography of the built environment, this reshaping involves organizational and structural changes.

2. Elsewhere, I have tried to show how these new inequalities in profit-making capacities of economic sectors, earnings capacities of households, and prices in upscale and downscale markets have contributed to the formation of informal economies in major cities of highly developed countries (Sassen, 1994). These informal economies negotiate between new economic trends and regulatory frameworks that were engendered in response to older economic conditions.

3. Much of the discussion around the formation of a single European market and financial system has raised the possibility, and even the need if it is to be competitive, of centralizing financial functions and capital in a limited number of cities rather than maintaining the current structure in which each country has a financial center.

4. The space economy of leading information industries raises a very specific question of control and governance. In these industries more so than in many others, a significant component of transactions and markets operates in an electronic space that is not subject to conventional jurisdictions. The question of control and regulation being engendered by the electronic and telecommunications side of this new space economy lies beyond much of the discussion about the shrinking role of the state in a global economy. Once transactions are embedded in these new technologies, speed alone creates problems of control that are new and cannot be handled through conventional state-centered or non-state forms of authority. The most familiar case is that of foreign currency markets, where volumes made possible by multiple transactions in a single day have left the existing institutional apparatus, notably the central banks, impotent to affect outcomes as they once were able to do. The space economy of these industries points to a reconfiguration of key parts of the governance debate. Besides the matter of globalization extending the economy beyond the reach of the state, it also is a matter of control that goes beyond the issue of interorganizational coordination that is at the heart of governance theory. Insofar as speed is one of the logics of the new information technologies, it does not always correspond with the logic of the economic institutional apparatus represented by finance and advanced services (Sassen, 1996, 1998, chap. 9). For an examination of how these new technologies could benefit sectors that lack power, see *Journal of Urban Technology* (1995).

5. For a brilliant historical account of other instances where this duality has not worked, see Taylor (1995). For an account focused on the impacts of digitalization, see *Futur Anterieur* (1995).

6. In the three decades after World War II, the period of the Pax Americana, economic internationalization had the effect of strengthening the interstate system. Leading economic sectors, especially manufacturing and raw materials extraction, were subject to international trade regimes. Individual states adjusted national economic policies to further this version of the world economy. Certain sectors did not fit comfortably under this largely trade-dominated interstate regime, and their escape produced the Euromarkets and offshore tax havens of the 1960s. The breakdown of some of these aspects of the Bretton Woods system during the mid-1970s created an international governance

void filled by multinationals and global financial markets. Inside the state, a further shift occurred away from agencies most closely tied to domestic social forces, as was the case during the Pax Americana, and toward agencies closest to the transnational process of consensus formation. By the late 1980s, concerted efforts to set up supranational and privatized governance mechanisms had come into being.

7. An emerging and significant issue in view of the spread of Western legal concepts is the critical examination of the philosophical premises concerning authorship and property that define the legal arena in the West (Coombe, 1993.)

8. Immigration is increasingly constructed as a devalued process insofar as it describes the entry of people from generally poorer, disadvantaged countries in search of the better lives that the receiving country can offer. It contains an implicit valorization of the receiving country and a devalorization of the sending country.

9. During the colonial era, the cities in the colonies probably were the most internationalized (King, 1989).

10. Such contestation and "slippage" can assume many different forms. Global mass culture homogenizes and is capable of absorbing an immense variety of local cultural elements. This process never is complete. The opposite is the case in my analysis of electronic manufacturing. It shows that employment in leading sectors no longer inevitably constitutes membership in a labor aristocracy. Thus, Third World women working in export processing zones are not empowered; capitalism can work through difference. Yet another case is that of "illegal" immigrants. Here, we see that national boundaries have the effect of creating and criminalizing difference. These types of differentiations are central to the formation of a world economic system (Wallerstein 1990).

11. Tokyo has several, mostly working class concentrations of legal and illegal immigrants coming from China, Bangladesh, Pakistan, and the Philippines. This is quite remarkable in view of Japan's legal and cultural closure to immigrants. Is this simply a function of poverty in those countries? By itself, it is not enough of an explanation given that they have long had poverty. I posit that the internationalization of the Japanese economy, including specific forms of investment in those countries and Japan's growing cultural influence there, have created bridges between those countries and Japan and have reduced the subjective distances to Japan (Cybriwski, 1991; Sassen, 1991, pp. 307-315).

12. The specific forms of the internationalization of capital over the past 20 years have mobilized people into migration streams. They have done so principally through the implantation of Western development strategies, from the replacement of small-holder agriculture with export-oriented commercial agriculture and export manufacturing to the Westernization of educational systems. At the same time, the administrative, commercial, and development networks of the former European empires and the newer forms these networks assumed under the Pax Americana (e.g., international direct foreign investment, export processing zones, wars for democracy) have created bridges for the flow of capital, information, and high-level personnel from the center to the periphery and for the flow of migrants from the periphery to the center (Sassen, 1996, chap. 3; 1998, chaps. 2-3).

13. A good description of largely unknown groups that have come to the forefront of urban life can be found in the volume based on the 1997 INURA conference on direct action groups (Wolff et al., 1998). See also Dunn (1994), Keil (1993), and *Social Justice* (1993).

REFERENCES

Abu-Lughod, J. L. (1995). Comparing Chicago, New York, and Los Angeles: Testing some world cities hypotheses. In P. L. Knox & P. J. Taylor (Eds.), *World cities in a world-system* (pp. 171-191). Cambridge, UK: Cambridge University Press.

Berner, E., & Korff, R. (1995). Globalization and local resistance: The creation of localities in Manila and Bangkok. *International Journal of Urban and Regional Research, 19*(2), 208-222.

Body-Gendrot, S. (1993). *Ville et violence.* Paris: Presses Universitaires de France.

Castells, M. (1989). *The informational city.* London: Blackwell.

Cohen, M. A., Ruble, B. A., Tulchin, J. S., & Garland, A. M. (Eds.). (1996). *Preparing for the urban future: Global pressures and local forces.* Washington, DC: Woodrow Wilson Center Press. (Distributed by Johns Hopkins University Press)

Coombe, R. J. (1993). The properties of culture and the politics of possessing identity: Native claims in the cultural appropriation controversy. *Canadian Journal of Law and Jurisprudence, 6,* 249-285.

Cybriwsky, R. (1991). *Tokyo: The changing profile of an urban giant.* London: Belhaven.

Dunn, S. (Ed.). (1994). *Managing divided cities.* Staffs, UK: Keele University Press.

Fainstein, S., Gordon, I., & Harloe, M. (1993). *Divided city: Economic restructuring and social change in London and New York.* London: Blackwell.

Friedmann, J. (1995). Where we stand: A decade of world city research. In P. L. Knox & P. J. Taylor (Eds.), *World cities in a world-system* (pp. 21-47). Cambridge, UK: Cambridge University Press.

Futur Antérieur. (1995). La ville-monde aujourd' hui: Entre virtualité et ancrage [special issue], *30-32.*

Graham, S., & Marvin, S. (1995). More than ducts and wires: Post-Fordism, cities, and utility systems. In P. Healey, S. Cameron, S. Davoudi, S. Graham, & A. Madani-Pour (Eds.), *Managing cities: The new urban context* (pp. 169-189). Chichester, UK: Wiley.

Hall, S. (1991). The local and the global: Globalization and ethnicity. In A. D. King (Ed.), *Current debates in art: History 3—Culture, globalization, and the world-system: Contemporary conditions for the representation of identity.* Binghamton: State University of New York, Department of Art and Art History.

Hitz, H., Keil, R., Lehrer, U., Ronneberger, K., Schmid, C., & Wolff, R. (Eds.). (1995). *Capatales fatales.* Zurich: Rotpunkt Verlag.

Journal of Urban Technology. (1995). Information technologies and inner-city communities [special issue], *3*(1).

Keil, R. (1993). *Weltstadt-stadt der welt: Internationalisierung und lokale politik in Los Angeles.* Münster, Germany: Westfaelisches Dampfboot.

King, A. D. (1989). Colonialism, urbanism, and the capitalist world-economy: An introduction. *International Journal of Urban and Regional Research, 13*(1), 1-18.

King, A. D. (Ed.). (1995). *Representing the city: Ethnicity, capital, and culture in the 21st century.* London: Macmillan.

Kowarick, L., & Campanario, M. (1986). São Paulo: The price of world city status. *Development and Change, 17*(1), 159-174.

Le Débat. (1994, Summer). Le nouveau Paris [special issue].

Martinotti, G. (1993). *Metropoli: La nuova morfologia sociale della citta.* Bologna, Italy: Il Mulino.

Noller, P., Prigge, W., & Ronneberger, K. (Eds.). (1994). *Stadt-Welt.* Frankfurt, Germany: Campus Verlag.

Peraldi, M., & Perrin, E. (Eds.). (1996). *Réseaux productifs et territoires urbains.* Toulouse, France: Presses Universitaires du Mirail.

Sachar, A. (1990). The global economy and world cities. In A. Sachar & S. Oberg (Eds.), *The world economy and the spatial organization of power* (pp. 149-160). Aldershot, UK: Avebury.

Sassen, S. (1991). *The global city: New York, London, Tokyo.* Princeton, NJ: Princeton University Press.

Sassen, S. (1994). *Cities in a world economy.* Thousand Oaks, CA: Pine Forge.

Sassen, S. (1996). *Losing control? Sovereignty in an age of globalization* (1995 Columbia University Leonard Hastings Schoff Memorial Lectures). New York: Columbia University Press.

Sassen, S. (1998). *Globalization and its discontents.* New York: New Press.

Simon, D. (1995). The world city hypothesis: Reflections from the periphery. In P. L. Knox & P. J. Taylor (Eds.), *World cities in a world system* (pp. 132-155). Cambridge, UK: Cambridge University Press.

Social Justice. (1993). Global crisis, local struggles [special issue], *20*(3/4).

Stren, R. (1996). The studies of cities: Popular perceptions, academic disciplines, and emerging agendas. In M. A. Cohen, B. A. Ruble, J. S. Tulchin, & A. M. Garland (Eds.), *Preparing for the urban future* (pp. 392-420). Washington, DC: Woodrow Wilson Center Press. (Distributed by Johns Hopkins University Press)

Taylor, P. J. (1995). World cities and territorial states: The rise and fall of their mutuality. In P. L. Knox & P. J. Taylor (Eds.), *World cities in a world-system* (pp. 48-62). Cambridge, UK: Cambridge University Press.

von Petz, U., & Schmals, K. M. (Eds.). (1992). *Metropole, weltstadt, global city: Neue formen der urbanisierung.* Dortmund, Germany: Universitat Dortmund, Dortmunder Beitrage zur Raumplanung.

Wallerstein, I. (1990). Culture as the ideological battleground of the modern world-system. In M. Featherstone, (Ed.), *Global culture: Nationalism, globalization, and modernity.* London: Sage.

Wissenschaft Forum. (1995). Global city: Zitadellen der internationalisierung [special issue], *12*(2).

Wolff, R., Schneider, A., Schmid, C., Klaus, P., Hofer, A., & Hitz, H. (1998). *Possible urban worlds.* Basel: Birkhauser Verlag.

6

Transnationalism and the City

MICHAEL PETER SMITH

Debates on "the global city" have a recognizable, if not formulaic, character—poised somewhere on a conceptual and epistemological borderland where positivism, structuralism, and essentialism meet. The tendency to focus this debate around positivist taxonomies, urban hierarchies constructed on the basis of these taxonomies, and empirical efforts to map or even formally model the "real" causes and consequences of global cities leads all sides to overlook the fact that the world cities debate takes place within a wider public discourse on "globalization," which is itself a contested political project advanced by powerful social forces, not something to be observed by the scientific tools of objective social scientists. The global cities discourse constitutes an effort to define the global city as an objectified and essentialized reality, a "thing" operating outside the social construction of meaning. The participants in this debate argue about which material conditions are attributes of global cities and which cities possess these attributes. The debate generates alternative positivist taxonomies of global city status.

Viewed in this light, the attributes taken as signs of economic globalization and global city status are most fully developed in the analyses offered by Friedmann (1986, 1995) and Sassen (1991). These include

AUTHOR'S NOTE: An earlier version of this chapter was prepared for delivery at a seminar on the dynamics of urban Asia sponsored by the Asia/Pacific Research Center, Stanford University, November 19, 1997. It draws in part on a commentary written for the *Urban Affairs Review* (February 1998) and on a more extended theoretical analysis of transnationalism written with my colleague, Luis Eduardo Guarnizo, for *Comparative Urban and Community Research* (1998). I thank Fred Block and John Dale for their helpful comments on this chapter and thank Luis Guarnizo for the collaborative relationship that has been central to my understanding of contemporary transnationalism.

five "nested" economic attributes: (a) internationalization of the world economy accompanied by increased economic transgression of national borders; (b) heightened capital mobility, the deployment of which is concentrated in a handful of global cities; (c) the shift from manufacturing to business and financial (or producer) services in major "core country" cities; (d) accordingly, the concentration within key global or world cities of global command-and-control functions coordinated by the growing producer service sectors found there; and (e) the hierarchical organization of these cities into a global system of cities whose purpose is the accumulation, control, and deployment of international capital. Only the interrelationships of these attributes within a positivist causal framework is subject to debate.

My own position on these issues is framed within the wider epistemological and ontological debate on social constructionism and the critique of ideology. The basic starting point of my argument is that no solid object known as the global city and appropriate for grounding urban research exists, only an endless interplay of differently articulated *networks, practices,* and *power relations* best deciphered by studying the agency of local, regional, national, and transnational actors who discursively and historically construct understandings of locality, transnationality, and globalization in different urban settings.

The global city is best thought of as a historical construct, not as a place or "object" consisting of essential properties that can be readily measured outside the process of meaning making. Debates about the essence of world cities are reminiscent of earlier debates over the "essence of place" (Massey, 1994, p. 111). In both debates, certain internal features of this or that place constitute its essential status. In the case of global cities, that status is the possession of a cumulative number of economic attributes that position the city within a stratified world hierarchy of cities. Instead of pursuing the quest for a hierarchy of nested cities arranged neatly in terms of their internal functions, it is more fruitful to assume a world of crisscrossing articulations of global and local, with sociocultural as well as political-economic relations operating both *outside* and *within* the borders and boundaries of today's urban centers. These partially overlapping and often contested networks of meaning are relations of power that link people, places, and processes to each other transnationally in disjointed rather than hierarchical patterns of interaction (see, e.g., Appadurai, 1990, 1996; Smith, 1997; Smith & Guarnizo, 1998).

Viewed from this perspective, all cities, rather than a handful of producer service centers, become germane to the comparative analysis of (a) the localization of global economic, sociocultural, and political flows; (b) the globalization of local socioeconomic, political, and cultural forces; and (c) the networks of social action connecting these flows and forces in transnational social space. These emergent transnational networks are human creations best understood as *sites of multicentered, if not decentered, agency,* in all of their overlapping untidiness. Thus, studying differences in the patterns of intersection of global and local flows at particular geographical conjunctures, in particular political spaces, is more fruitful than cataloging the economic similarities of hierarchically organized financial centers viewed as constructions of a single agent—multinational capital.

The "new urban politics" uncovered by this move is a disjointed terrain of global media flows, transnational migrant networks, state-centered actors that side with and oppose global actors, local and global growth machines and green movements, multilocational entrepreneurs, and multilateral political institutions, all colluding and colliding with each other ad infinitum. The urban future following from this contested process of "place making" is far less predictable but far more interesting than the grand narrative of global capital swallowing local political elites and pushing powerless people around that inevitably seems to follow from the global cities model. Rather than viewing global cities as central expressions of the global accumulation of capital, *all* cities can be viewed in the fullness of their particular linkages with the worlds outside their boundaries. Their "urban" specificity, in short, becomes a matter of discerning "the particularity of the social interactions which intersect at that location, and of what people make of them, in their interpretations and in their lives" (Massey, 1994, p. 117).

This brings me to the second part of my argument. The global cities literature needs to be situated historically as part of a far larger academic and public discourse on the meaning of globalization. Recent work by Drainville (1998), McMichael (1996), and others (Mander & Goldsmith, 1996; Smith, 1997) argues that globalization, like its binary local diversity, needs to be understood as a historical construct rather than as an objective economic process operating behind our backs. Whose historical construct is it, anyway?

According to McMichael (1996), the grand narrative of economic globalization has been advanced most forcefully not by academics but rather by an emergent international monetarist regime, a set of institutional actors who have instituted a political offensive against developmental states and institutions. The globally oriented institutions spearheading this offensive were established "under the auspices of the 1980s debt crisis" (p. 25) to advance the monetarist agenda of global efficiency and financial credibility against the nationally oriented institutions of developing countries. This global political project has produced, in turn, a series of struggles over the meaning of "the local," as cities and other localities, via their political, economic, and cultural actors and institutions, seek either to find a niche within the new globalist public philosophy or to resist pressures to "globalize," that is, to practice fiscal austerity and conform to monetarist principles and policies.

The origins of the ideology of globalization are historically specific. They constitute efforts by powerful social forces to replace the developmentalist institutional framework that existed from the 1960s to the 1980s, itself premised on modernization theory, with a new mode of economic integration of cities and states to world market principles. These principles have been posited by their advocates as inevitable by-products or ruling logics of the material condition of globalization "on the ground" rather than, as they are, the social constructs of historically specific social interests. McMichael (1996) and Drainville (1998) identify these interests as including the heads of international agencies, state managers who have embraced neoliberalism, transnational corporate financial institutions, various academic ideologues, and (most especially) "the managers of newly empowered multilateral institutions like the IMF [International Monetary Fund], the World Bank, and the World Trade Organization" (McMichael, 1996, p. 28). This neoliberal regime of "global governance" (Drainville, 1998) is viewed as an incipient global ruling class whose efforts to achieve global economic management can be thwarted only when globalization itself is recognized as a historically specific and contested project of social actors and agents rather than as an inevitable condition of contemporary existence.

McMichael (1996) and others clearly overestimate the coherence of this global project and give scant attention to the effectiveness of oppositional forces (Sikkink, 1993; Smith, 1994). Thus, although it is not possible to agree entirely with their assessment of the global future, it is nonetheless surprising that none of the global cities scholars, whose politics surely would oppose global governance on neoliberal principles, seems to have given much thought to the question of whether their research agenda and its objective findings implicitly support that project by legitimating the "reality" of global cities as part and parcel of the "objective processes" of economic globalization. This question brings to the fore two basically political questions: "The global city: Whose social construct is it anyway?" and "If not the global city, then what?"

Understanding Transnational Urbanism

Because all representations of cities and urbanization processes are social constructs, why is the terrain of transnationalism a more fruitful optic for guiding urban research than the global cities framework that I have just critiqued? At first glance, transnationalism seems as caught up in the politics of representation that characterizes the current post-structuralist moment in the social sciences as the global cities framework is mired in an unhappy marriage of structuralism and positivism. For example, the current expansion of transnational capital, mass media, and international migration on a global scale has provoked a spate of discourses on the "crisis of the nation-state." In both the social sciences and cultural studies, a core theme has been the penetration of national cultures and political systems by global and local driving forces. The nation-state is seen as weakened "from above" by transnational capital, global media, and emergent supranational political institutions (Cvetkovich & Kellner, 1997; Drainville, 1998). "From below," it is envisaged as facing the decentering local resistances of the informal economy, ethnic nationalism, and grassroots activism (Mahler, 1998; Portes, 1995, 1996). These developments sometimes are represented in celebratory terms as harbingers of global market rationality from above or of liberating practices from below such as the cultural hybridity brought to postmodern urban life by transnational migration. In more pessimistic readings, these developments are represented as preludes to a new form of capitalist modernization that will convert the entire planet to global consumerism. In most of these representations, transnationalism from above and below is represented as something profoundly "new."

Yet, transnational connections forged by international trade, global migration, cross-border political movements, and cultural interpenetrations are by no means entirely new phenomena. Each of these flows has a very long history, be it the sojourns of Marco Polo, the economic alliances of city-states, the mass migrations of the early 20th century, or the political actions leading to the unification of Italy. Nevertheless, at the present historical moment, four contemporary processes contributing to the formation of transnational networks of social action do appear to be new. These processes appear to be driving both the turn to transnationalism as an urban research focus and the contentious character of the debate on the role of cities in a putatively "post-national" global cultural economy (Appadurai, 1996). They include (a) the *discursive repositioning of cities* in relation to nation-states in the ongoing debate on the meaning of globalization; (b) the emergence of *cross-national, political, and institutional networks* that deploy the discourses of decolonization, human rights, and other universalist tropes to advance the interests of heretofore marginalized others; (c) the facilitation of transnational social ties by *new technological developments* that have *widened access* to the means of transnational travel, communication, and ways of being in the world; and, following from these, (d) the *spatial reconfiguration of social networks from below* that facilitate the reproduction of migration, business practices, cultural beliefs, and political agency.

Starting from this multidimensional conception of contemporary transnationalism, I wish to explore four themes central to the emergence of transnational urbanism as a research field: (a) the sociospatial agency of transnational social, economic, and political networks; (b) the need to study what has come to be termed the *translocality;* (c) the continuing significance of the nation-state as a player in the social construction of transnational urbanism; and (d) the need for *comparative* urban analysis of emergent transnational networks viewed as articulations taking place within, between, and across the local sites that heretofore have been the domains of urban research.

■ The Agency of Transnational Networks

Transnational sociocultural and political-economic agency occur at multiple spatial scales (Smith, 1994). The social construction of transnational networks should be treated as the result of separate, sometimes parallel, and sometimes competing projects at all levels of analysis within the putatively global system, from the global governance agenda of international organizations and multinational corporations to the most local social practices by which transnational networks are constructed. At the global level of analysis, the specific multilateral collectivities identified previously by McMichael (1996) and Drainville (1998) seek to construct a global neoliberal contextual space, a "new world order," to regulate transnational flows of capital, trade, people, and culture. In the process, they supplant the disintegrating nationally managed regimes of Keynesianism and Fordism (Drainville, 1998). At the most local level, specific collectivities—local households, kin networks, elite factions, and other emergent local formations—actively pursue strategies such as transnational migration (Goldring, 1998; Matthei & Smith, 1998; Smith, 1998), transnational social movements (Guarnizo & Smith, 1998; Mahler, 1998; Sikkink, 1993; Smith, 1994), and transnational economic or cultural entrepreneurship (Portes, 1995, 1996; Schein, 1998) to sustain or transform resources, including cultural resources, in the face of the neoliberal storm.

Transnational practices connect social networks located in more than one national territory. This does not mean that such practices are inherently transgressive bearers of new social subjectivities. Rather, transnational political, economic, and sociocultural practices are embodied in historically specific, culturally constituted social relations; that is, they are networks of meaning, established between particular spatially and temporally situated social actors. Thus, the "local" dimension of transnationalism needs to be considered carefully. Although once viewed in social theory as the site of stasis or settlement, one of the main contributions of postmodern ethnography and critical theory has been the redefinition of "the local" as a dynamic source of alternative cosmopolitanisms and contestation (Robbins, 1993; Schein, 1998; Smith, 1992). In light of this more dynamic view of locality, it is important to consider the shifting boundaries of transnational urbanism as reflected in the enabling or constraining sociocultural networks found in particular sites, very often large cities.

This disruption of the binary of local stasis versus global dynamism highlights the representation of "the urban" in transnational discourse and vice versa. In the version of transnational urbanism I am advancing in this chapter, cities are not to be viewed as empty containers of transnational articulations. Transnational flows, such as capital investments, migration patterns, and IMF policies, are not simply imposed on cities from the outside. Rather, the local often reconstitutes the meaning of globalization. For example, because local cultural understandings mediate and, therefore, mutually constitute the meaning of investment, it matters significantly whether Hong Kong capital investment networks are forged vis-à-vis Shenzhen or Vancouver (cf. Smart & Smart, 1998, to Mitchell, 1996). Likewise, because local cultural practices and political understandings mediate global power relations, it matters whether IMF austerity policies are implemented in San Jose (Costa Rica) or Mexico City (cf. Tardanico, 1995, to Barkin, Ortiz, & Rosen, 1997). Because local social constructions of immigrants may differ from one city to the next, it matters whether migrants from Latin America and the Pacific Rim move to New York, Miami, Los Angeles, or San Francisco, to say nothing of whether they move to Asian or European metropoles (cf. Mahler, 1996, to Smith & Tarallo, 1993; cf. also the various case studies in Nonini & Ong, 1997; Smith & Guarnizo, 1998).

My intention here is not to essentialize local culture. The social production of local cultural practices in New York or Los Angeles is a very heterogeneous enterprise. Because cities, local states, and community formations are not bounded, self-contained, coherent entities but rather multiple and even contradictory, and because no necessary correlation exists between any place and the cultural meanings informing the practices that occur there, researchers must leave open the analytical questions of how and when the local mediates the global. The local, national, and transnational connections that go into the making of transnational urbanism are mutually constitutive.

There is a rapidly growing body of literature on transnational connections (see, e.g., the case studies in Basch, Schiller, & Blanc, 1994; Hannerz, 1996; Nonini & Ong, 1997; Smith & Guarnizo, 1998). This literature grounds the study of transnationalism in the crisscrossing networks of social practice that collude and collide in particular cities at particular times. The historically specific patterns of politics, culture, and economic life found in particular locales significantly mediate the transnational flows of people, resources, ideas, and information. Any given city receiving particular transnational economic, political, or cultural flows provides a specific configuration of potential opportunities and constraints (e.g., labor market conditions, investment opportunities, popular and official perceptions of a migrant group, the presence or absence of other political activists) into which migrants, investors, or political and cultural brokers enter. Thus, the play of agency operates differently from place to place and even within the same place at different times. Because of power differentials among the various networks interacting in particular places at particular times, the local social context of these interactions is in constant flux. Yet at any given moment, the confluence of various inward flows and specific urban and regional settings shapes not only the likelihood of generating, maintaining, or forsaking transnational ties but also the very character of the ties that can be forged. An unraveling of the complexity of transnational urbanism calls attention to the continuing significance of cities as grounded sites of meaning making.

The Rise of Translocalities

Recent research on transnationalism illustrates that the specific social space in which transnational actions take place is not merely local but often "translocal" (i.e., local to local). Translocal relations are constituted within historically and geographically specific points of origin and destination established by transnational migrants, investors, political activists, and sociocultural entrepreneurs. They form a multifaceted connection that links together transnational actors, the localities to which they direct flows, and their points of origin. The points of origin implicated in transnational urbanism may be cities such as Hong Kong, where investment capitalists direct investments to other locales on the Pacific Rim; rural villages in Mexico, where households send transmigrants to New York or Los Angeles; smaller cities such as Oakland, where a hotel and restaurant workers union "local" initiates transnational, multi-ethnic, working class organizing campaigns; or centers of political ferment such as Amsterdam and the San Francisco Bay Area, where activists deploy and reterritorialize various transnational human rights discourses and practices. The social relations forged by transnational networks link such sending locations with receiving cities in complex ways, generating *translocal* discursive and spatial practices that may reconfigure and even transform relations of power.

Consider as a prime example a recent study by Smart and Smart (1998) of the investment practices of entrepreneurial Hong Kong capitalists in the Shenzhen region of mainland China. The Smarts' work investigates the historically particular forging of translocal relations that mediate the pattern of global investment. Their study reveals a particular pattern of "situated ethnicity" as the basis for constructing translocal network solidarity and exclusion that differs markedly from the types of translocal ties forged by corporate Chinese capitalists from Hong Kong who are channeling investment into Vancouver (Mitchell, 1993, 1996). These studies illustrate that in penetrating different cities and regions in the world economy, transnational capitalist factions from Hong Kong are not entirely free agents. Indeed, they have to justify their activities within prevailing local cultural understandings. The entrepreneurial Hong Kong capitalists studied by Smart and Smart foreground their "Chinese-ness" in Shenzhen, whereas the corporate capitalists from Hong Kong in Vancouver accommodate to a different setting by downplaying their Chinese-ness and foregrounding their capitalist economic position within a dominant multicultural public discourse. Far from erasing local identifications and meaning systems, transnationalism actually relies on them to sustain transnational ties.

Several of the contributions to the recent collection *Ungrounded Empires,* edited by Ong and Nonini (1997), nicely illustrate the dynamic character of the sorts of local and translocal social ties that go into the making of transnational urbanism. Ong and Nonini have assembled fascinating studies of the forging of translocal social relations that are central to our understanding of both modern Chinese transnationalism and contemporary Pacific Rim urbanization. Following James Clifford's injunction to consider "travel" as a necessary supplement to "settlement" in conducting ethnographic research (Clifford, 1992), the contributors lucidly explore the social networks and practices formed by the movements of Taiwanese entrepreneurs into and out of South China (Hsing, 1997); the sojourns of Malaysian ethnic Chinese transnational migrant workers in Japanese cities (Nonini, 1997); and the discourses of national, transnational, and pan-Asian identity played out among ethnic Chinese minorities living in Bangkok and metropolitan Manila (Blanc, 1997).

These studies of transnational urbanism in Asian cities are complemented by a spate of recent research in migration studies on the rise of transnational communities that are sustained by social ties forged by transmigrants moving between small villages in Latin America and large U.S. cities (Goldring, 1998; Mahler, 1998; Rodriguez, 1995; Smith, 1998). By *transnational communities,* these researchers mean *translocality-based structures of cultural production and social reproduction.* Such transnational sociocultural structures are sustained by social networks in migration and their attendant modes of social organization. Such networks include translocal "hometown" associations in Los Angeles, New York, and Houston; the transmission of economic remittances from income earned in transnational cities to sustain families, promote community development projects, and transform local power and status relations in villages of origin; and the reconfiguration of local cultural festivals and celebrations in both sites to signify translocality. Translocal connections are sustained as well by more indirect technological means of transportation and communication—jet airplanes, satellite dishes, courier services, telephones, faxes, e-mail—now available to facilitate the production of transnational social ties.

These processes of constructing translocality cannot be readily reduced to the economistic rubric of "household reproduction." Indeed, the everyday practices of transmigrants, including the remittances they control, involve not only household survival but also the reinscription of status positions that enhance the political influence of transmigrants in sending locales (Goldring, 1998; Smith, 1998). The resources mobilized by transmigrants also have been shown to reconfigure community power relations (Nagengast & Kearney, 1990; Smith, 1998) in sending and receiving locales. Transnational bodily movement is not a movement across a neutral space but rather an encounter involving an often contentious meeting ground of different urban racial hierarchies and different local social relations of gender. These translocal zones of experience have opened up new spaces for the transgression of racial and gendered boundaries (see, e.g., the case of Dominican urban transmigrant women and blacks in New York and Madrid in Sørensen, 1998). Louisa Schein's recent case study of the invention of Hmong transnational ties with the Miao ethnic minority in China shows that it is even possible to completely reinvent one's ethnic origins by the production, diffusion, and consumption of culturally oriented ethnic videos laden with geographical images and cultural icons (Schein, 1998). By these means, Hmong refugees from Laos living in small and large cities in California, Minnesota, and elsewhere are constructing a myth of cultural origins linked not to Laos but rather to the Miao regions of China.

As this example suggests, the social construction of what Benedict Anderson calls "imagined communities" (Anderson, 1983/1991) inextricably links discourses of localism, nationalism, and transnationalism in very complex ways. When the studies by Schein (1998) and Smart and Smart (1998) are compared, we find that political elites of the local states in different Chinese regions forge links and construct a cultural sense of "we-ness" with U.S.-based Hmong cultural brokers and Hong Kong-based entrepreneurs, respectively, that bypass national party loyalties and the ideology of the Chinese state. Institutional actors representing the Chinese state are eager to attract foreign remittances and investment; therefore, they tolerate these translocal ties within China's borders. Yet, they remain watchful, worried about the risks of ethnic separatism and the erosion of the ruling party's control of local politics.

Likewise, when processes of translocality construction that link villages in Mexico to U.S. cities are compared to the translocal Hong Kong–Shenzhen connection, we find quite different translocal social constructions of identity and affiliation. Unlike the Mexican transmigrants, who are concerned with using transnational connections to reconfigure power and status relations within their villages of origin and maintain a reconfigured hometown identity (Goldring, 1998; Smith, 1998), the small capitalists of Hong Kong studied by Smart and Smart (1998) carefully avoid establishing economic ties in their own villages of origin in China for fear of the "excess" (i.e., non-business related) expectations and demands that might be thrust on them. Their basis for building transnational economic and social relations is situated ethnicity (i.e., a constructed, thoroughly modern, transnational Chinese-ness) rather than hometown loyalty (for alternative readings of Chinese transnationalism, see Nonini & Ong, 1997; Ong, 1997).

Questioning the Post-National Discourse

Does this way of conceiving transnational connections mean, as some (e.g., Appadurai, 1996) have claimed, that transnational networks from above and below are now ushering in a new period of weakened nationalism, a post-national global cultural economy? Does it mean, as global cities devotees suggest, that the nation-state faces inexorable decline? There are several reasons to treat these claims with some skepticism.

First, in the present period of heightened transnational migration, many nation-states that have experienced substantial out-migration are entering into a process of actively promoting transnational reincorporation of migrants into their state-centered projects. Why is this so? Global economic restructuring and the repositioning of states, especially less industrialized ones, in the world economy have increased the economic dependency of these countries on foreign investment. Political elites and managerial strata in these societies have found that as emigration to advanced capitalist countries has increased, the monetary transfers provided by transmigrant investors have made substantial contributions to their national and urban economies (Lessinger, 1992). Likewise, transnational household remittances have promoted social stability and new forms of local community development (Kearney, 1991; Mahler, 1996; Smith, 1994). Thus, a growing dependence on transmigrants' stable remittances has prompted sending states to incorporate their "nationals" abroad into both their national markets and their national polities by a variety of measures including naming "honorary ambassadors" from among transmigrant entrepreneurs in the hope that they will promote national interests vis-à-vis receiving countries; subsidizing transnational migrant hometown and "home state" associations (Goldring, 1998; Mahler, 1998; Smith, 1998); creating formal channels for communicating with these constituencies across national borders (Schiller & Fouron, 1998; Guarnizo, 1999; Nagengast & Kearney, 1990); passing dual citizenship laws; and even, in the bizarre case of the state apparatus in El Salvador, providing free legal assistance to political refugees so that they may obtain asylum in the United States on the grounds that they have been persecuted by the state that is paying their legal expenses. Far from withering away in the epoch of postnationalism, sending states once presumed to be peripheral are promoting the reproduction of transnational subjects. In the process, they are reinventing their own role in the new world order.

Second, the agents of receiving states remain relevant actors. Although weakened by the neoliberal assault in their capacity to promote social welfare policies within their borders, states still monopolize the legitimate means of coercive power to police their borders. It might even be argued that the very uncertainties of national identity posed by transnational penetration of national boundaries from above and below has heightened the symbolic and material consequences of policing borders. The rise of restrictive immigration laws in receiving societies speaks amply to this issue. Thus, it is problematic to represent as a "deterritorialization of the state" the expansion of the reach of states of origin beyond their own national territorial jurisdictions into other state and urban formations. Rather, when politicians from Mexico, for example, come to campaign and proselytize to Mexican transnational migrants living in Los Angeles or Fresno (Nagengast & Kearney, 1990), or when businesspeople from China act in various ways to open or maintain markets in U.S. or European cities, their influence still is exercised in a particular territorial domain, formally controlled by institutional actors of the receiving state. The juridical construction of transnational social formations by these state-centered actors often is one that denies their "globality" and reterritorializes their meaning as a "boundary penetration," that is, as a transgression of the state's own jurisdiction. The recent political controversy in the United States concerning the penetration of Asian money into U.S. national political campaigns is a case in point. It suggests that the social actions of political elites who rule nation-states are not merely reactions to but also partially constitutive of the scope and meaning of transnationalism within their territories. Thus, state-initiated discourses remain important to the reinscription of nationalist ideologies and national subjects in the face of transnationalism both from above and from below.

There is a third reason for skepticism concerning the post-national character of the cultural and political economy of cities in transnational times. Although cities might welcome the transnational investment flows that contribute to their economic development, scant evidence exists that the local context of reception for transnational migrants from below has been equally receptive. Rather than being welcomed for their talents, resourcefulness, and/or hybridity, transmigrants often are stigmatized and stereotyped as signs of poverty, difference, and urban decline. Thus, rather than engendering a public discourse on their contributions to the vitality of urban life in transnational cities, the presence of transmigrants in U.S. cities has reignited nationalist discourses on the "disuniting" of America (Schlesinger, 1992).

Paradoxically, the expansion of transnational migration has resulted in outbursts of entrenched, essentialist nationalism in both sending and receiving locales. In receiving cities and states, movements aimed at recuperating and reifying mythical national identities are expanding as a way in which to eliminate the penetration of alien "others." States of origin, on the other hand, are reessentializing their national identities and extending them to their nationals abroad as a way in which to maintain their loyalty and flow of resources "back home." By granting them dual citizenship, these states are encouraging transmigrants' instrumental accommodation to receiving societies while simultaneously inhibiting their cultural assimilation and thereby promoting the preservation of their own national cultures. This, in turn, further fuels nativist sentiments in receiving cities and states.

Undoubtedly, transnational practices cut across the politically instituted boundaries of cities and states. These transnational actions are nonetheless affected by the policies and practices of territorially based local and national states and communities. Yet in addition to the representations of elites that control national territories and local states, the power-knowledge systems shaping imagined communities at today's urban moment also are shaped by visions of trade, travel, migration, locality, translocality, and diaspora emanating from the meaning making of transnational networks. Thus, transnational urbanism is mutually constituted by the shifting interplay of local, national, and transnational relations of power and meaning.

Toward a Transnational Urban Studies

Clearly, there is a need to expand the study of transnational urbanism to encompass the scope of transnational processes, as well as to focus future urban research on the local and translocal specificities of various transnational sociospatial practices. Traditional methods for studying people in cities—ethnography, life histories, historical case studies—must be rethought. The challenge is to develop an optic and a language capable of representing the complexity of transnational connections and the shifting spatial scales at which agency takes place.

Given this complexity, a fruitful approach for research on transnational urbanism would start with an analysis of networks situated in the social space of the city and with an awareness that the social space being analyzed might best be understood as a translocality, a place where institutions interact with structural and instrumental processes in the formation of power, meaning, and identities. By contrast, starting from the global level and deducing urban outcomes from global developments, as in the global cities framework, often leads to overgeneralization and produces the self-fulfilling grand theories that have been the postmodern object of derision. This is particularly problematic when scholars (Giddens, 1991; Harvey, 1990; Jameson, 1984) become so wrapped up in the theoretical elegance of their formulations (e.g., late capitalism, time-space distantiation or compression) that they ignore questions of how the world "out there" is imagined, socially constructed, and lived.

An equally problematic pitfall would be to begin and end analysis of transnational urbanism at a purely local level. In privileging local knowledge (Geertz, 1983), researchers might develop a solipsistic tunnel vision that altogether fails to connect human intentions to social networks and historical change. Situating the study of transnational urbanism at the level of particular cities viewed as translocalities avoids this pitfall. Specifically, we need to ask how transnational networks operate and how principles of trust and solidarity are constructed across national territories as compared to those that are entirely locally based and maintained. What place-making discourses and practices hold transnational networks together? How are social connectedness and control organized across borders to guarantee commitment? How do transnational relations interact with local power structures including class, gender, and racial hierarchies? With what effects? How does translocality affect the sociocultural basis supporting transnational relations and ties?

The task of reconfiguring urban research for studying translocalities presents serious challenges and offers new opportunities for creative scholarship. Most of the current research on transnational networks has been required by the scope of transnational relations to pursue a multi-locational research strategy that crisscrosses urban, national, cultural, and institutional boundaries (see the studies in Ong & Nonini, 1997; Smith & Guarnizo, 1998). For example, Schein's (1998) inventive deployment of unorthodox ethnographic methods moves back and forth between text and context, observation and participation, and localities in the United States and in China, acting out her self-described role as an ethnographic nomad. As Clifford (1992) suggests, the study of "traveling cultures" requires traveling researchers. As increasing numbers of formerly locality-based social networks, grassroots movements, and entrepreneurial activities extend across national boundaries, becoming binational (if not multinational) in spatial scale, urban research needs to be literally "re-placed" from the local to the translocal and transnational scales.

A current limitation of existing knowledge about transnational networks, translocalities, and transnational urbanism is the dearth of comparative urban studies. In my view, future urban research ought to focus considerable attention on comparatively analyzing diverse cases of *transnational network formation* and *translocality construction.* Such studies might take different forms, offering insights concerning different modes of transnational urbanism.

Three particularly useful modes of comparison come to mind. First, we might compare the practices of the same transnational network in different cities (whether the network is a migrant group, an investment network, or a participating component of a transnational social movement) to determine the effect of the local on the transnational. A second approach would be to compare and contrast transnational practices undertaken by different social networks in the same city in terms of the different local effects of different modes of transnational social organization and the different impacts that the social organization of the particular city has on the networks. Finally, we might compare the consequences of neoliberal economic policies in different cities where they have been "localized" (e.g., Bangkok and Singapore or Mexico City and San Jose [Costa Rica]) to focus the interplay of the global and the local on the production of new spaces of domination, accommodation, and resistance.

In pursuing this project, it would be a mistake to conceptualize transnational urbanism as an easily quantifiable "thing" such that a city may be conceived as being more or less transnational. This would lead to the same tendency to create taxonomies and nested hierarchies that has limited the usefulness of the global cities approach. Transnationalism is neither a thing nor a continuum of events that can be easily quantified; rather, it is a complex process involving macro- and microdynamics. Thus, it is more fruitful to study in comparative historical perspective the effects that different transnational networks have on different cities, power structures, identities, and social organizations as well as the effects of the latter on the former.

In sum, the optic of transnational urbanism is preferable to the global cities approach to urban research for several related reasons. It is an agency-oriented perspective that enables us to see how globalization is constructed by the historically specific social practices that constitute transnational networks. The post-national assumptions of the global cities framework are replaced by an approach in which the nation-state is given its due as an institutional actor that is implicated in the process of forming and reconstituting transnational ties. Ordinary people are viewed as creative actors involved in the social construction of transnational urbanism by the social networks they form rather than being viewed as passive objects propelled to global cities by underlying logics. Local forms of socioeconomic organization, political culture, and identity formation are taken into account without being essentialized, or even erased, as in some variants of the global cities model. The rise of translocalities and the meaning-making practices that occur in these new social spaces are shown to be central to the social organization of transnational urbanism.

Global cities research is limited by its focus on the economic organization of a handful of command-and-control centers, its desire to construct a panoptic hierarchy of cities, and its neglect of culture as a product and producer of difference. The study of transnational urbanism expands our vision to encompass the full range of cities implicated in the political-economic and sociocultural practices of transnational migrants, entrepreneurs, political activists, and institutions. It moves us beyond a global cities approach that attributes too much power to capital, too much uniformity to urbanization, too much hierarchy to interurban relations, and too little autonomy to culture. In its place, transnational urbanism offers an approach capable of grasping the complex differences intersecting within and between the world's cities. These differences are socially produced by the networks and circuits of social interaction intersecting in particular cities and in particular lives. To envision cities as sites of transnational urbanism, transnational networks must be localized, their connections must be translocalized, and their practices must be historicized. This is a central task for future urban research.

REFERENCES

Anderson, B. (1991). *Imagined communities: Reflections on the growth and spread of nationalism.* London: Verso. (Originally published in 1983)

Appadurai, A. (1990). Disjuncture and difference in the global cultural economy. *Public Culture, 2*(2), 1-24.

Appadurai, A. (1996). *Modernity at large: Cultural dimensions of globalization.* Minneapolis: University of Minnesota Press.

Barkin, D., Ortiz, I., & Rosen, F. (1997). Globalization and resistance: The remaking of Mexico. *NACLA Report on the Americas, 30*(4), 14-27.

Basch, L., Schiller, N. G., & Blanc, C. S. (1994). *Nations unbound: Transnational projects, postcolonial predicaments, and the deterritorialized nation-state.* New York: Gordon and Breach.

Blanc, C. S. (1997). The thoroughly modern "Asian": Capital, culture, and nation in Thailand and the Philippines. In A. Ong & D. Nonini (Eds.), *Ungrounded empires: The cultural politics of modern Chinese transnationalism* (pp. 261-286). New York: Routledge.

Clifford, J. (1992). Traveling cultures. In L. Grossberg, C. Nelson, & P. Treichler (Eds.), *Cultural studies* (pp. 96-116). New York: Routledge.

Cvetkovich, A., & Kellner, D. (Eds.). (1997). *Articulating the global and the local: Globalization and cultural studies.* Boulder, CO: Westview.

Drainville, A. (1998). The fetishism of global civil society. In M. P. Smith & L. E. Guarnizo (Eds.), *Transnationalism from below* (pp. 35-63). New Brunswick, NJ: Transaction.

Friedmann, J. (1986). The world city hypothesis. *Development and Change, 17*(1), 69-84.

Friedmann, J. (1995). Where we stand: A decade of world city research. In P. L. Knox & P. J. Taylor (Eds.), *World cities in a world system* (pp. 21-47). Cambridge, UK: Cambridge University Press.

Geertz, C. (1983). *Local knowledge: Further essays in interpretive anthropology.* New York: Basic Books.

Giddens, A. (1991). *Modernity and self-identity: Self and society in the late modern age.* Stanford, CA: Stanford University Press.

Goldring, L. (1998). The power of status in transnational social fields. In M. P. Smith & L. E. Guarnizo (Eds.), *Transnationalism from below* (pp. 165-195). New Brunswick, NJ: Transaction.

Guarnizo, L. E. (1999). The rise of transnational social formations. *Political Power and Social Theory, 12,* 45-94.

Guarnizo, L. E., & Smith, M. P. (1998). The locations of transnationalism. In M. P. Smith & L. E. Guarnizo (Eds.), *Transnationalism from below* (pp. 3-34). New Brunswick, NJ: Transaction.

Hannerz, U. (1996). *Transnational connections.* London: Routledge.

Harvey, D. (1990). *The condition of postmodernity.* Oxford, UK: Blackwell.

Hsing, Y.-T. (1997). Building *Guanzxi* across the straits: Taiwanese capital and local Chinese bureaucrats. In A. Ong & D. Nonini (Eds.), *Ungrounded empires: The cultural politics of modern Chinese transnationalism* (pp. 143-164). New York: Routledge.

Jameson, F. (1984). Postmodernism, or the cultural logic of late capitalism. *New Left Review, 146,* 53-92.

Kearney, M. (1991). Borders and boundaries of state and self at the end of empire. *Journal of Historical Sociology, 4,* 52-74.

Lessinger, J. (1992). Investing or going home? A transnational strategy among Indian immigrants in the United States. In N. G. Schiller, L. Basch, & C. S. Blanc (Eds.), *Toward a transnational perspective on migration: Race, class, ethnicity, and nationalism reconsidered* (pp. 53-80). New York: New York Academy of Sciences.

Mahler, S. J. (1996). *American dreaming: Immigrant life on the margins.* Princeton, NJ: Princeton University Press.

Mahler, S. J. (1998). Theoretical and empirical contributions towards a research agenda for transnationalism. In M. P. Smith & L. E. Guarnizo (Eds.), *Transnationalism from below* (pp. 64-100). New Brunswick, NJ: Transaction.

Mander, J., & Goldsmith, E. (Eds.). (1996). *The case against the global economy.* San Francisco: Sierra Club Books.

Massey, D. (1994). Double articulation: A place in the world. In A. Bammer (Ed.), *Displacements: Cultural identities in question* (pp. 111-121). Bloomington: Indiana University Press.

Matthei, L., & Smith, D. (1998). "Belizean boyz 'n the hood"? Garifuna labor migration and transnational identity. In M. P. Smith & L. E. Guarnizo (Eds.), *Transnationalism from below* (pp. 270-290). New Brunswick, NJ: Transaction.

McMichael, P. (1996). Globalization: Myths and realities. *Rural Sociology, 61*(1), 25-55.

Mitchell, K. (1993). Multiculturalism, or the united colors of capitalism? *Antipode, 25,* 263-294.

Mitchell, K. (1996). In whose interest? Transnational capital and the production of multiculturalism in Canada. In R. Wilson & W. Dissanayake (Eds.), *Global/local: Cultural production and the transnational imaginary* (pp. 219-254). Durham, NC: Duke University Press.

Nagengast, C., & Kearney, M. (1990). Mixtec ethnicity: Social identity, political consciousness, and political activism. *Latin American Research Review, 25*(2), 61-91.

Nonini, D. (1997). Shifting identities, positioned imaginaries: Transnational traversals and reversals by Malaysian Chinese. In A. Ong & D. Nonini (Eds.), *Ungrounded empires: The cultural politics of modern Chinese transnationalism* (pp. 203-227). New York: Routledge.

Nonini, D., & Ong, A. (1997). Chinese transnationalism as an alternative modernity. In A. Ong & D. Nonini (Eds.), *Ungrounded empires: The cultural politics of modern Chinese transnationalism* (pp. 3-38). New York: Routledge.

Ong, A. (1997). Chinese modernities: Narratives of nation and of capitalism. In A. Ong & D. Nonini (Eds.), *Ungrounded empires: The cultural politics of modern Chinese transnationalism* (pp. 171-203). New York: Routledge.

Ong, A., & Nonini, D. (Eds.). (1997). *Ungrounded empires: The cultural politics of modern Chinese transnationalism.* New York: Routledge.

Portes, A. (1995). *Transnational communities: Their emergence and significance in the contemporary world system.* Working Papers Series, No. 16, Program in Comparative and International Development, Department of Sociology, Johns Hopkins University.

Portes, A. (1996, March-April). Global villagers: The rise of transnational communities. *The American Prospect,* pp. 74-77.

Robbins, B. (1993). Comparative cosmopolitanisms. In B. Robbins (Ed.), *Secular vocations: Intellectuals, professionalism, culture* (pp. 180-211). London: Verso.

Rodriguez, N. (1995). The real "new world order": The globalization of racial and ethnic relations in the late twentieth century. In M. P. Smith & J. R. Feagin (Eds.), *The bubbling cauldron: Race, ethnicity, and the urban crisis* (pp. 211-225). Minneapolis: University of Minnesota Press.

Sassen, S. (1991). *The global city: New York, London, Tokyo.* Princeton, NJ: Princeton University Press.

Schein, L. (1998). Forged transnationality and oppositional cosmopolitanism. In M. P. Smith & L. E. Guarnizo (Eds.), *Transnationalism from below* (pp. 291-313). New Brunswick, NJ: Transaction.

Schiller, N. G., & Fouron, G. (1998). Transnational lives and national identities: The identity politics of Haitian immigrants. In M. P. Smith & L. E. Guarnizo (Eds.), *Transnationalism from below* (pp. 130-163). New Brunswick, NJ: Transaction.

Schlesinger, A. (1992). *The disuniting of America.* New York: Norton.

Sikkink, K. (1993). Human rights, principled issue networks, and sovereignty in Latin America. *International Organization, 47,* 411-441.

Smart, A., & Smart, J. (1998). Transnational social networks and negotiated identities in interactions between Hong Kong and China. In M. P. Smith & L. E. Guarnizo (Eds.), *Transnationalism from below* (pp. 103-129). New Brunswick, NJ: Transaction.

Smith, M. P. (1992). Postmodernism, urban ethnography, and the new social space of ethnic identity. *Theory and Society, 21,* 493-531.

Smith, M. P. (1994). Can you imagine? Transnational migration and the globalization of grassroots politics. *Social Text, 39,* 15-33.

Smith, M. P. (1997). Looking for globality in Los Angeles. In A. Cvetkovich & D. Kellner (Eds.), *Articulating the global and the local: Globalization and cultural studies* (pp. 55-71). Boulder, CO: Westview.

Smith, M. P., & Guarnizo, L. E. (Eds.). (1998). *Transnationalism from below.* New Brunswick, NJ: Transaction.

Smith, M. P., & Tarallo, B. (1993). *California's changing faces: New immigrant survival strategies and state policy.* Berkeley: California Policy Seminar.

Smith, R. (1998). Transnational localities: Community, technology, and the politics of membership within the context of Mexico and U.S. migration. In M. P. Smith & L. E. Guarnizo (Eds.), *Transnationalism from below* (pp. 196-239). New Brunswick, NJ: Transaction.

Sørensen, N. (1998). Narrating identity across Dominican worlds. In M. P. Smith & L. E. Guarnizo (Eds.), *Transnationalism from below* (pp. 241-269). New Brunswick, NJ: Transaction.

Tardanico, R. (1995). Economic crisis and structural adjustment: The changing labor market of San José, Costa Rica. In M. P. Smith (Ed.), *After modernism: Global restructuring and the changing boundaries of city life* (pp. 70-104). New Brunswick, NJ: Transaction.

7

The Urban World Through a South African Prism

ALAN MABIN

> *We should be grateful to the poor of the cities of the South for what they teach us, not only about our past, but about the future that awaits us if we do not recognize our common fate and act accordingly.*
>
> —Jeremy Seabrook (1996, p. 301)

South African cities have mostly been considered in their exceptionalism, with specific focus on that country's (no longer) legally institutionalized racism and the patterns of segregation that resulted. The possibility exists, however, that as South African cities change after apartheid, they will allow examination of paths of urban development that speak to segregation elsewhere in the new world spatial order of cities. How we conceptualize urban South Africa might well enable a wider conceptualization of the city.

Segregation, of course, is not the only process to which investigation of South African cities can contribute. Northern cities as well as southern cities reveal multiple patterns (e.g., persistent economic uncertainty, growing violence and fear) to which the urbanisms of South Africa can speak. If we break from thinking of any particular urban model (e.g., Chicago, Los Angeles) as "the future," then we can enrich our urban concepts to cope with the dizzying pace and myriad places of urban life. The common trends of contemporary urbanisms, however,

AUTHOR'S NOTE: Many of the ideas in this chapter were developed in conversations with or stimulated by Sue Parnell (University of Cape Town), Bob Beauregard (New School for Social Research), and Abdou Maliq Simone (University of Witwatersrand). I also acknowledge comments from participants in conferences and seminars in Midrand (Society of South African Geographers, 1997), London (Institute of Commonwealth Studies, 1997), and La Rochelle (Ministère Française de l'Equipement, 1998).

are difficult to see when we are overwhelmed by particular circumstances. Unless we break with that problem, one that has bedeviled many schools and epochs of urban thought, we will arrive at rather linear ideas of change. In this context, South African cities can help us "to recognize our common fate and act accordingly."

If we persist in portraying South African cities as having derived from a colonial background, however, we will not achieve this goal. Thinking of the cities as somehow consisting of First World and Third World elements is weak logic for constructing the nature of the cities. The range of features of South African (and other) cities is present in both sets.

The idea of cities as First World and Third World is hardly unique to South Africa. Michael Dear offers the image of the postmodern city as such a combination. For Dear (1995, p. 40), as for other well-known researchers, Los Angeles is the image of the future, and that future contains the extremes that characterize cities of the North and those of the South. But the developing polarizations to which this image supposedly refers have been obvious to observers of cities in other parts of the world for decades. In South Africa, the idea of society as a combination of First World and Third World very frequently lapses into racism or at least is used to legitimate persistent separations including those between new elites and the expanding poor population. This discourse goes back to modernization economics (with its dual economy) and blocks an understanding of urban society, however segregated it might appear to be, as a phenomenon with many intersections, dependencies, and overlaps among its apparently separate elements.

If we misunderstand the nature of the cities, we will get policy wrong, wasting resources on targets that are not amenable to the attention that our incorrect conceptualization suggests would be appropriate. In fact, the end of apartheid planning, with its opening of distant vistas of different urban worlds, creates the possibility of getting it all wrong. Fear concentrates thought and pushes one back toward a more intellectual realm. From a scholarly perspective, addressing our conceptualization of South African cities is a necessary exercise. Our current concepts are derivative, often shallow, sometimes self-centered.

This chapter concentrates on four predominant features of our existing conceptualization of urban South Africa: (a) segregation, (b) linear concepts of urbanization, (c) a persistent equation between industrialization and urbanization, and (d) violence. It uses the changing nature of South African urbanism as a kaleidoscope whose images recur in other urban contexts. The limits of these concepts are profound. Thus, our rethinking must draw on bodies of ideas produced in other urban contexts. I hope to show how developing new ideas about South African cities can enrich the global urban literature and not merely re-map the narrow (but obviously rich) field of South African urban studies (Parnell, 1997).

From Segregation to Postcolonialism and Postmodernity

We think of South African cities as unique because of the impact of apartheid. These cities have mostly been considered in their exceptionalism, that is, as products of legally institutionalized racism with resulting patterns of segregation. A. J. Christopher performs a magnificent service in showing how much of the segregationist thinking that shaped these cities is merely something that they share with the unfortunate history of colonial urbanisms (Christopher, 1992). Still, we need to go beyond the simple characterization of apartheid cities as a variant on a broad colonial theme. For example, increasing degrees of segregation characterize many cities in other parts of the world. What is odd and noteworthy about South Africa is not the fact of apartheid's impact on urban space but rather that a concerted effort was mounted to overcome segregation and associated phenomena and that this effort is contrary to global urban trends (Mabin, 1995).

After World War II, explicit group oppression rapidly became less acceptable globally, thereby rendering apartheid an isolated (not to mention highly oppressive) approach to policy. Nonetheless, cities everywhere continued, and in some ways have intensified, the polarization and marginalization of identifiable social groups. This contradiction can be seen in sharp relief in the cities of southern Africa, particularly South Africa, where decolonization, the end of apartheid, globalization, the discrediting of socialism, and structural adjustment create conditions in which the struggle to overcome social injustice in the city is confronted by ever stronger social and spatial segregation and/or fragmentation.

One consequence is that old forms of urbanism have been, or are being, superseded. In large cities everywhere, the old predominance of single-centeredness is giving way to diversity of activity and function, at least to multi-centeredness, and is altering the ways in which people relate to urban activities, their patterns of movement, and their experience of urban life. The forms of segregation of those who enjoy differential access to the good things of the city are shifting. The ways in which people use urban space—the meaning of urban space itself—are changing. Examples can be found in heightened competition for resources, enterprise survival, and sustenance. A striking physical illustration is located in the vicinity between Pretoria and Johannesburg, the original "centers" of which are 60 kilometers (approximately 35 miles) apart. Until recently a sparsely populated rural domain, the space between the two is now the scene of suburban commercial and industrial growth every bit as dramatic as that in many Asian and American locales. Having achieved the end of apartheid, the previously excluded can occupy in every sense the old dominant core of the city (as in Johannesburg), but as they do so (creating new urbanisms in the process), the centers of opportunity slip away toward the suburban and exurban zones, once again out of reach of those left behind (and this time more elusive in their very diversity and fluidity).

In the dominant view, South African cities after apartheid must be moving toward something "normal." But toward what? Obviously, toward market-based forms of segregation, as Peter Marcuse argues (Marcuse, 1995). Instead of rigid and racist "group area" rules, mortgage redlining excludes needy borrowers, house prices filter classes, and the economic geography of renting creates new forms of ghettos. On the other hand, these market-based changes do not necessarily point toward, say, American types of segregation. The possession of real power (at first political and, as time passes, increasingly economic) on the part of previously excluded groups generates (at least for now) racially mixed elite neighborhoods, even in the oldest bourgeois districts. Think of the scenes of President Nelson Mandela's 80th birthday in 1998, viewed globally on television, celebrated in his private home in an older, very high-income section of Johannesburg. That area lacks evidence of "white flight," a finding that is widely replicated.

Indeed, a peculiarity of South African cities is the very limited extent of "tipping" in the American sense, that is, of near total change in the color of residents after a threshold proportion is reached. Although some neighborhoods have indeed tipped from white to black, such as the well-known Hillbrow (Morris, 1999), there are deep economic reasons for limited geographical change. In general, enclaves of extensive, inexpensive, rental accommodation have been entirely altered, but not anywhere else. Instead, segregation has developed new forms that are almost aspatial, related mainly to access to economic activity and differential mobility. In all South African cities, on the same block or at least in the same neighborhood, one can observe "the close proximity of diverse ways of life, sometimes intersecting but more often existing in a high density of parallel urban worlds" (Simone, 1998).

The demise of apartheid has had a great deal to do with complex polarizations in urban populations (Crankshaw, 1997). In South Africa, as elsewhere, new extremes of polarization exist, some of which are related to the accumulation of the increasing array of electronic commodities (e.g., video recorders, portable personal computers, CD players, mobile phones, digital cameras) that continually increase the distance between those who have or use them and those who do not. Those who do not also are increasingly deprived of access to other supports, particularly as the state reduces its already minimal provision of collective services.

These processes can be starkly seen in the South African context. Just where it might be expected that the defined disabilities of the mass of "ordinary" citizens would be removed (and in some respects *are* removed), it becomes evident that those who previously shared racially defined disadvantage can readily be divided by a process of rapid and radical polarization between those who gain access to new positions and those who do not. The lines along which those polarizations occur bear examination as well. They speak of many different systems of ascription of status, some rooted in lineage, some in political patronage, and some in other forms of entrepreneurialism.

These fragmentations—economic and social, sometimes political, and with geographical expression—are perhaps what led Beauregard and Haila (1987) to argue that urban form is a combination of modern and postmodern elements and always will be. Postmodern work, perhaps, is precisely that which tries to grapple with these combined elements. Perhaps we could add to this the combination of colonial and postcolonial elements in the sense described by Jane Jacobs in *Edge of Empire* (Jacobs, 1996). The conditions of postmodernity and postcoloniality suggest fruitful paths of exploration in South African cities, and they offer representations that might help to reconceptualize urban segregation. The possibility also exists that as South African cities change after apartheid, they will allow examination of paths of urban development that speak to segregation elsewhere.

Circular Migration Rather Than Linear Change

Cities of the South cannot be conceptualized adequately as social entities without situating them firmly in the context of massive, continuing, long-term population movement in and out of the city. The meanings of urbanism in this type of political economy need to be explored.

We generally imagine (at least outside of North America and Western Europe) that our cities inevitably are growing on the basis of rural-to-urban migration, and we think of this as taking place because the cities offer something that rural life does not. The consequences are that we construct a linear model of urbanization, and all sorts of people and politicians fall into thinking that rural development will change the course of urbanward migration. Instead of this linear view, I offer the concept of persistent circular migration (Mabin, 1990) as a means of understanding what is going on and of evaluating urban and rural policy.

This notion has become familiar. It suggests that rather than an inevitability attaching to removals from rural to urban space, big proportions of populations (national and frequently international) can engage very complex patterns of movement between rural and urban spaces (and categories in between). Multiple bases for often fragmented but sometimes recomposed households make for circulation of diverse individuals among places. The circularity means that when the circle is unbroken, urban populations are intimately linked to rural populations. When centripetal forces break the integrity of the circle, people end up in all sorts of places—a further form of fragmentation—which has especially important effects for identity and politics, as Mahmoud Mamdani argues so eloquently (Mamdani, 1996, esp. chaps. 6-7). The circles and breaks are incomplete in South Africa and probably increasingly so in many other cases.

We must make a clean intellectual break with the persistent idea that South Africa's shifting patterns of circular migration are unusual and that apartheid created the complexities of migration. Of course, it contributed. Circular migration was built on apartheid foundations and reshaped them, but apartheid was not the cause. Apartheid's abolition simply increases the puzzle of relations between urban and nonurban life. Urbanization has to be discussed, investigated, and understood not in terms of linear change in relationships between rural and urban.

The particular urban-rural nexus might be what is distinct about poorer cities, although they show plenty of variety. It certainly has a special place in understanding African cities, particularly those south of the equator (e.g., the differences between South African and Brazilian cities). Yet, the exploration of urban-rural connections (i.e., of permanent-temporary dialectics and, more substantively, of household formation and dissolution) has been little actualized in urban research.

On the policy side, attempts to redistribute rural land increasingly exacerbate the support of continuing poverty if Sam Moyo's work is to be believed. Moyo (1986) argues that land reform can create the basis of more deeply persistent circular migration through guaranteeing multiple points of passage rather than providing sustainable rural livelihoods. Evidence from Zimbabwe supports this thesis, and there is Mexican evidence as well. Here is a startling example of the contradictory effects of policy development that misconstrues its urban context and directs our attention to the need "to link the rural and the urban in ways that have not yet been done" (Mamdani, 1996, p. 296).

Just as the origin groups (e.g., Indian, Russian, Nigerian) from which cab drivers are drawn in New York change at an accelerating pace, South African cities demonstrate bewildering social diversity that is recomposing their divisions of labor and integrations of cultures. In the global view, all cities move toward cosmopolis whether or not the planners understand (Sandercock, 1998). The vitality of social links and their significance to the negotiation of place, rights, and income in the city are apparent, although as yet in an opaque way, because South African cities await research of the type that has informed understanding of forms of West African urbanism (Diouf, 1996; Mbembe & Roitman, 1996). Close parallels exist between the latter's images of urbanized populations that continue to operate in locations marginal to the urban system and actual urban economies, creating the "high density of parallel urban worlds" to which reference already has been made.

Industrialization and Urbanization

Just as disturbing as the linear notions of urbanization criticized in the preceding section is the dominant borrowing from the tenacious global idea that modern urbanism is intimately linked with industrialization. If anything, South African cities illustrate how much divergence there is between the course of industrialism and the course of urbanism. Its cities simply do not show long-term correlations in industrial workforces and total populations, something they have in common with most cities of the South. This distance is growing in wealthier parts of the world as well and has long existed in Gareth Stedman-Jones's portrayal of 19th-century London (Stedman-Jones, 1971).

The official conception of "the urban" in South Africa, during the years of industrial employment growth from segregation to apartheid, linked industrialization closely to urbanization. That linkage meant that those whose lives were not well-connected to industry were excluded from the cities. Another consequence was the push from urban to rural at just the time that state support for land clearances created great pressure for urban space. This approach created, in so many places, a city good for elites. Similar, but not necessarily official, exclusions now take place.

More fluid concepts of the urban note that urban has not been simply equated to industry. A large portion of South Africa's "real urban economy" is being developed in the interstices between the formal and the informal, the traditional and the provisional, highly normative codes and improvised social arrangements (Simone, 1998).

Sundering the urban-industry connection, the drive no longer is to attract the industries that in the past had provided jobs, paid taxes, and generally supported the city. During the era of globalization, the imperative is not about what industries cities can attract but rather where cities can successfully place themselves in global arenas. As Borja (1996) indicates, the ability of cities to act on information regarding international markets, the flexibility of commercial and productive structures, and the capacity to enter networks of various dimensions and complexities constitute the criteria for viability over and beyond the factors of geographical location, prior positions within national or international economies, and accumulated capital or natural resources. Urbanization is no more equal to industry now than it has been at any other time.

So, urbanism takes forms totally devoid of industry, whether seeking to place itself with local reference or with global orientation. In, at the edges, and sometimes well beyond, the established cities are the non-centers of a growing new urbanism. In South Africa, places such as Hammanskraal, north of Pretoria, demonstrate different forms of property development, trade, and social relationships linked to trade, property, and production. Unregulated, it is a new and wild city. Such places are not unlike Kariobangi (at the edge of Nairobi) and remind one of the energy of "irregular" urbanization at the edges of São Paulo and many other Latin American cities. In some places, urbanism increasingly comes down to these forms—West Africa is replete with examples—and each lurch of booming urban economies in Asia exposes more evidence of similar relationships.

■ Violence: A New Urbanism

Just as urbanism and industry seem more remote from one another, citizenship and civility, those other quintessentially urban qualities, seem less and less complete even in the midst of a democratization such as South Africa's. The issue is violence, and violence (and its variant property crimes) permeates any description of South African cities. Like so much else, violence is neither distinctive to the cities of South Africa nor confined to one or another sphere of society, although the loss of a state monopoly on violence has been nowhere more acutely evident than in South Africa. One constant is that urban violence is directed at, or at least unleashed against, marginalized people in cities across the globe.

The gated community responses of Los Angeles, São Paulo, and Johannesburg (Caldeira, 1996; Davis, 1990) attest to this level of fear and violence. These communities represent the re-creation of the compound, much like those built to house and control workers at the gold mines of southern Africa, but this time with the poor walled out instead of in.

With crime and violence intensified, new types of cities might appear, cities that are not merely theaters of visible accumulation but rather are theaters of invisible accumulation. The cities of Africa south of the Sahara are increasingly marginal to the globalizing world. From the perspectives of people on their streets, the globalizing world often is peripheral to their lives as well. Planning in these cities attempts to adopt models of strategic management and to apply them in the mold of Western Europe or Southeast Asia. Yet, the nature of the state and of urbanism as forms of social organization frustrates aspects of these planning approaches and fosters other planning approaches. "Civil society" and something perhaps better described as "uncivil society" (e.g., violence, crime, warlordism, patronage) alternately undermine and support these efforts with unpredictable results. Thus, we need to find ways in which to conceptualize African cities in the world, challenging some of the analyses associated with globalization and the consequent prescriptions. Again, South African cities might shed light.

Our understanding of these issues can be cast in the form of a tension between local (internal) and global (international) forces shaping urbanism. Spatial form provides excellent clues. On the one hand, spatial forms associated with deep degrees of polarization (e.g., informalization of crime, violence, and space) suggest the significance of local actors in giving shape to urban life. On the other hand, a wider global terrain can construct spaces that the nation-state fails to reach. In this sense, it is possible for poorer cities to become central to crime. Their potential as refuge shelters—the squatter camps of the world—and the impact on polarization in wealthier cities can be both imagined and, perhaps in the near future, illuminated. Contemporary urban contestation, of which violence is an ineluctable part, produces a situation in which things can happen very quickly.

Conclusion

South African cities can help us, in Africa and globally, "to recognize our common fate and act accordingly." South African cities share an emerging global urban characteristic, one with which the four categories of discussion in this chapter have tried to grapple and one that demands new ways of thinking. The common ground is massive wealth production by fewer and fewer of the overall (urban) population (emphasizing the delinking of industry and urbanism) and constricted consumption of its fruits by even fewer and ever more exalted people. The distribution system fails to match the production system. We see these processes at work in South Africa perhaps more clearly than observers based in different countries because they appear to have accelerated under the post-1994 democratic government. Democracy and openness to the world have thrown so much of the urban into relief and into uncertainty and even anxiety.

It is as though we had to wait for uncertainty in Los Angeles to make apparent what already was obvious in the large cities of Brazil, Indonesia, and Nigeria—"that all urban place-making bets are off; we are engaged, knowingly or otherwise, in the search for new ways of creating cities" (Dear, 1995, p. 44). Observers in the South might describe what has happened in "postmodern urbanism" differently; a short account would simply be that the cities of the North have joined those of the South in persistent economic uncertainty, multiculturalism to a degree that changes the character of preceding (colonial and white-dominated) urbanism, growing violence and fear, and accelerating polarization in social and economic life. Still, the northern cities have not yet joined the southern cities in the large (if uneven) *proportion* of the population that lives in complete insecurity. In this sense, the cities of South Africa show potential futures and not only possible pasts of the cities of the North (Harvey, 1997). Creative maladjustment (as Martin Luther King Jr. put it during the 1960s) to the dominance of Western linear models of urban futures, the apparent inevitability of polarization, and increasing deprivation for an increasing number seem necessary. If these are the directions of change, then it seems obvious that urban policy and planning in southern contexts must largely be about social questions. In the North, it has been easier to avoid that conclusion.

The South African urban environment is increasingly one in which it is difficult to ascertain just what social practices, alliances, and knowledge can be mobilized that will be sufficient to meet the professed development aspirations of its citizens (Simone, 1998). Therefore, the deepening of South African urban studies offers possibilities of enormous scholarly value to global urban research. Reconceptualizing urban South Africa offers insights valuable to the wider conceptualization of the city.

Lastly, and beyond the scholarly domain, deepening these new conceptions and applying them to policy formation is essential if urban environments are to experience positive development in the next century. On this depends an urban renaissance that touches the vast urban population now rising in the cities of the South, specifically the lives of the excluded everywhere.

REFERENCES

Beauregard, R. A., & Haila, A. (1997). The unavoidable incompleteness of the city. *American Behavioral Scientist, 41,* 327-341.

Borja, J. (1996). Cities: New roles and forms of governing. In M. Cohen, B. Aruble, J. S. Tulchan, & A. M. Garland (Eds.). (1996). *Preparing for the urban future* (pp. 242-263). Baltimore, MD: Johns Hopkins University Press.

Caldeira, T. (1996). Fortified enclaves: The new urban segregation. *Public Culture, 8,* 303-328.

Christopher, A. J. (1992). Urban segregation levels in the British overseas empire and its successors in the twentieth century. *Transactions of the Institute of British Geographers, 17*(1), 95-107.

Crankshaw, O. (1997). *Race, class, and the changing division of labour under apartheid.* London: Routledge.

Davis, M. (1990). *City of quartz.* London: Verso.

Dear, M. (1995). Prolegomena to a postmodern urbanism. In P. Healey, S. Cameron, S. Davoudi, S. Graham, & A. Madani-Pour (Eds.), *Managing cities* (pp. 27-44). Chichester, UK: Wiley.

Diouf, M. (1996). Urban youth and Senegalese politics: Dakar, 1988-1994. *Public Culture, 8,* 225-249.

Harvey, D. (1997). *Justice, nature, and the geography of difference.* Oxford, UK: Blackwell.

Jacobs, J. M. (1996). *Edge of empire.* London: Routledge.

Mabin, A. (1990). Limits of urban transition models in understanding the dynamics of urbanization in South Africa. *Development South Africa, 7,* 311-322.

Mabin, A. (1995). On the problems and prospects of overcoming segregation and fragmentation in southern Africa's cities in the postmodern era. In S. Watson & K. Gibson (Eds.), *Postmodern cities and spaces* (pp. 187-198). Oxford, UK: Blackwell.

Mamdani, M. (1996). *Citizenship and subject: Contemporary Africa and the legacy of late colonialism.* Cape Town, South Africa: David Phillip.

Marcuse, P. (1995). *Race, space, and class: The unique and global in South Africa* (Southern Africa in Transition). Occasional Paper Series 4, Department of Sociology, Witwatersrand University.

Mbembe, A., & Roitman, J. (1996). Figures of the subject in times of crisis. In P. Yaeger (Ed.), *The geography of identity* (pp. 153-186). Ann Arbor: University of Michigan Press.

Morris, A. (1999). *Bleakness and light: Inner city transition in Johannesburg.* Johannesburg, South Africa: Witwatersrand University Press.

Moyo, S. (1986). The land question. In I. Mandaza (Ed.), *Zimbabwe: The political economy of transition 1980-1986* (pp. 165-202). Dakar, South Africa: Codesria.

Parnell, S. (1997). South Africa cities: Pespectives from the ivory tower of urban studies. *Urban Studies, 34,* 891-906.

Sandercock, L. (1998). *Towards cosmopolis.* Chichester, UK: Wiley.

Seabrook, J. (1996). *In the cities of the South: Scenes from a developing world.* London: Verso.

Simone, A. M. (1998). *Critique of the urban development framework.* Discussion paper, Isandia Institute, Capetown, South Africa.

Stedman-Jones, G. (1971). *Outcast London.* New York: Pantheon.

Part III

Civic Engagement

8

A City for Whom? Transients and Public Life in the Second-Generation Metropolis

GUIDO MARTINOTTI

Jacques-Albert Berthélemy had the sinecure of the archives of the Maltese Templar Order in Paris. This enabled him to become one of the many well-to-do and self-contented "mice in the cheese" who fed on the rotting body of the Ancien Régime. His privileges included taking possession of the premises in the *petite tour* of the order's Grandpriory in the heart of Paris. He remodeled the old premises with patience and dedication, particularly the furniture, creating a highly sophisticated urban mansion from which he managed his urbane style of life. Much to his disgrace, he had spun his parasitic cocoon on a dangerous spot. It was successively torn down but still was one of Paris's historic places.

During the night of August 10, 1792, the hand of fate knocked at Berthélemy's door, which also was the door of the famous dungeon, *la Tour du Temple.* His nice apartment was commandeered to become the temporary abode of the family of Citizen Capet. Without too much ado, Berthélemy was thrown into the courtyard in his nightgown and slippers with a few other personal belongings, his precious furniture confiscated with the rest of his libertine's nest, and his odyssey began. Obsessed with the idea of recovering his treasured furniture, he spent the following months roving through the Terror in search of his possessions. Armed with the courage of the demented, the distraught aristocrat filed

AUTHOR'S NOTE: I give credit to the European Foundation for the Improvement of Living and Working Conditions and to the European Union for permission to reprint and revise parts of the English version of my 1997 report to them.

plea after plea to get back his "meubles" with some of the cruelest and most sanguinary heads of the Terror, brushing sleeves with the likes of Saint Just, Hebert, or Chaumette, totally oblivious to the revolution raging around him.

I tell this anecdote not only to promote Victor Meyer-Eckardt's delightful historical novel, *I mobili del signor Berthélemy* (Meyer-Eckardt, 1946), but also because it perfectly epitomizes the situation of the modern urbanite. The citizen of the contemporary urban world is a little like Berthélemy. A worldwide revolution is evicting him from his sophisticated home, and in the midst of this revolution, he is desperately looking for the possessions of which he already has been stripped. While walking or driving in the new urban land, he continually longs for a disappearing city or muses about the "city of the future" without recognizing that such a city probably already is there. The task of the scholar is to help him find his way by making him aware of the revolution in which he is living and by giving him the right lenses to look at an urban reality already vastly different from the one imprinted in his heart or on his mind.

Although the changes undoubtedly are deep and radical, one must not expect the old city to crumble as in a disaster movie. The city, says Anthony Giddens, "displays a specious continuity with preexisting social orders" (Giddens, 1990, p. 6). Changes occur continuously, mostly by stalagmitical accretion, and we feel their immediate consequences on our skin when they affect daily urban practices, becoming increasingly unbearable and unrenounceable at the same time. The pace of change, however fast, still is gauged by historical standards. Where phenomena proceed in highly mixed ways, the end results become perceivable only in discrete quantum leaps.

When we observe one of those general trends that Emile Durkheim, apropos of the movement from country to town, called "un courant d'opinion, une poussée collective" (Durkheim, 1898), it is reasonable to think that these currents anticipate or reflect the reactions of the "collective soul" to a change of deep *structural* nature. A number of signs suggest that such a process is affecting the contemporary city. On the one hand, we notice the interruption and even the inversion of urbanization trends of secular breadth. On the other hand, the interest of scholars, local elites, and the public for the urban question grows disproportionately. In great cities, the mundane problems of daily life, from traffic congestion to the quality of air and water, are the object of uninterrupted attention and discussion. Equally crucial appear the problems of control over the social environment—widespread episodes of violence and criminality (and the related anxieties), the growing difficulties of providing collective urban services, and the various forms of local financial crises. The quality of air, the very *medium* absolutely essential for the physical survival of living organisms, is being monitored in real time in a growing number of urban areas, from Milano to Los Angeles.

At the same time, governing elites of all major and minor centers are increasingly enthralled by the idea of *city marketing,*[1] namely by the mix of competitive localization advantages that any given city can boast. Witness the strenuous fights between cities to attract important events such as the Olympic Games, soccer world championships, festivals, jubilees, and exhibitions. The main cities of the world get together in *clubs* and *lobbies* while a growing number of daily newspapers have specific sections dedicated to metropolitan issues.

The enticing images of the new technologies blend with the subtle anxieties of daily life and with the morbid visions of an incipient urban Middle Age, à la Gotham City. These visions have become highly intermeshed with the adventist mood brought about by the approaching end of the millennium. Every generation, particularly in our change-conscious era, wants to be at the watershed of history, and this century has provided a bounty of symbolic turning points. Among the plethora of millennial signs, a particularly significant one tends to be forgotten. Around the turn of the century, more than one of every two inhabitants on the planet has come to live in the social and physical context created some 60 to 100 centuries back. Precisely because of the growing number of persons, this context will be vastly different not only from the original but also from that of a few decades ago.

In this chapter, I sketch the conceptual tools needed to stimulate a growing awareness of the processes involved and their social consequences. I add my contribution to an increasing body of literature on contemporary urbanization[2] and do so in three steps. First, I highlight some of the unresolved questions that arise from the observation of current urban trends. Second, I suggest some answers based on a heuristic scheme proposed in my previous work (Martinotti, 1996, 1992/1997a, 1997b). Third, I identify the consequences of the emerging social morphology for the governance of the contemporary urban world, particularly for the fate of public spaces. I conclude by raising basic questions concerning the ways in which we observe the city.

The Ambiguities of Urban Development

The Exploding Metropolis

During the 1960s, the rhythm of urban growth led public opinion and scholars alike to formulate more or less apocalyptic visions of modern societies destined to be devoured by monstrous and impossible-to-run cities. In the wake of a solid academic tradition that since the 1930s has analyzed the development of large conurbations (Gottman, 1961) and of the interest shown by the great international organizations for growth, the theme of the large city, seen as a "world problem" (Hicks, 1974), was impressed on the public with a lasting effect, due to survive even the contrary statistical evidence that later surfaced. The term *megalopolis* (literally "great city"), coined by Jean Gottman to define the new form of urban settlements, progressively lost its original descriptive meaning to acquire a new one, strongly evocative of a quasi-pathological phenomenon, a cancer creeping over the world.

On the other hand, city growth appeared as the inevitable and unquestionable product of capitalistic economic development and of human progress *tout court.* Hope Tisdale wrote in 1942, "Urbanization is a process of concentration of population and implies a movement from a state of minor concentration to a state of greater concentration" (Tisdale, 1942, quoted in Berry, 1976, p. 17), offering a definition unchallenged for decades.[3]

> Urbanization proceeds in two ways: with the multiplication of concentration points and with the growth of the size of each concentration point. Occasionally, or in certain areas, urbanization can slow down or even turn back, but the tendency is *inherent in society* and proceeds until put in check by adverse conditions. (Tisdale, 1942, p. 311, ephasis added)

Even Marxist thought, although constantly alert about the social ills of great cities, treated them as largely the result of the capitalist mode of production and therefore destined to be eliminated by the wider transformation of society. In the wake of a strong intellectual position given to the problem by Marx himself, more wary of rural idiocy than of *malaise urbaine,* the Western Marxist tradition always has been concerned with overcoming the so-called "city-country contradiction" by extending urban advantages to the countryside rather than with the ultimate Pol Pot dystopia of erasing the city.[4] Supporters of capitalistic development joined its critics in believing urban growth to be an unavoidable and positive phenomenon, even if eventually to be corrected in its worst aspects.

The Slump in Urban Growth

As is well known, during the second half of the 1970s and the following decade, as well as in almost all the countries with advanced economies, urban growth underwent an abrupt and unforeseen slowdown. Before the end of the 1970s, urban development appeared universally bound to never end. During the following decade, the slowing down became apparent. At the same time, the much popularized images of the crisis of all great industrial metropolises—from Glasgow to New York, Detroit, St. Louis, and Turin—set the stage for an inversion of previous apocalyptic prophecies. During the preceding decades, cities were depicted as "exploding" (*Fortune* Magazine Editors, 1958) or as some sort of poisonous growth enveloping the planet. Now, the doomsayers started to talk about the "death of the city," deurbanization, and even an implausible "return to the countryside." True, the death of the city had been heralded in various occasions in the past,[5] but this time the theoretical elaborations seemed to be supported by incontestable evidence. From the beginning of the industrial revolution, the population had grown following the rule that "the larger the center, the faster its growth." After 1971, this rule began to be challenged. In the intercensual decade 1971-1981, the United States saw its nonmetropolitan population grow faster than its metropolitan population. Similar trends were recorded in many European countries, although with varying degrees and in successive waves. These results were relayed to the public with great emphasis by the media and by scholars who shared the excitement of the discovery of such downturns.

The Paradox of Deurbanization

As always during periods of deep structural transformation, the conceptual apparatus is first to be challenged, compelling us to improperly use old names and categories for new phenomena. We expend words about the city of the future without actually looking at what is happening. To give an idea of how misleading the lack of proper concepts can be, it suffices to recall how dominant the idea of deurbanization has been in scientific literature and even more in the popular one, starting from its first signs during the early 1980s. Statistical data on the loss of population in the core cities of large conurbations as well as new developments in the diffusion of home electronics were interpreted as ominous signs of the death of the city, to be replaced by diffused clusters of teleworkers.

It was a mistake. Trends observed so far in most of the advanced economies indicate that cities are not disappearing but rather are undergoing a profound transformation, the full consequences of which still are to be fathomed. In large-scale social change, the old and the new are highly mixed in reality as well as in our minds, and it is difficult to sever one from the other. If we do not recast rather radically our thinking about cities, then it will be impossible to even approximately forecast future developments despite the flood of symposia, special issues, and research projects. Our concepts continue to be patterned on a model of the city that already has largely disappeared, while the new metropolis is developing before our eyes as an incomprehensible freak. It is all too easy to list the theoretically possible conditions that will make cities livable places in the future, from clean air to social peace and well-organized and abundant social services. Without a clear idea of what is actually happening in the urban world, however, such a list will be an exercise in futility.

My main contention is that in current urban analyses, many functions are considered, but the *residential* one is greatly overstated. Simple evidence of this lies in the fact that most statistics about cities are based on *residential patterns* and residential *units of observation.* It is quite evident, however, that the new form of urban morphology is largely the product of the progressive individuation of several populations gravitating toward metropolitan centers and, in particular, of four populations that are increasingly differentiating one from the other.

■ An Analytical Framework

The First-Generation Metropolis

In the traditional town, on which all the current thinking about urban life still is largely molded, the inhabitants or the population *living* in the city coincided almost totally with the population *working* in the city. City limits encompassed both of these populations in one territory. For centuries, and until very recently, this space was encircled by walls and neatly separated from the rest of the land. The industrial revolution did not greatly change this situation; production of goods in the secondary sector requires mostly the shifting of raw materials, manufactured goods, and capital, whereas workers and entrepreneurs remain largely concentrated in urban areas. Only the 20th century brought about a radical change.

The early metropolitan development that took place from the 1920s onward in the United States and after World War II in Europe can essentially be seen as a growing differentiation of two populations: the *inhabitants* and the *workers. Commuting* is the main component of this process, namely the the most exemplary and widespread circadian experience of the urban dweller of the 20th century, but not the only one. If we look at the shape of the city during the first part of this century, we easily see how much commuting imprinted on the urban scene in terms of large-scale infrastructure, the creation of new and distinctive residential areas, and radical changes in the old centers. The result is what I call *first-generation* (or *early*) *metropolis,* characterized by a core city and by large surrounding areas, mostly based on functional urban regions or *commuting basins.*

Cities always depended on exchanges with their *hinterland* and always will do so. This is not a novelty. What is new is the nature and the range of the exchange. The well-known argument that it took an hour to cross Athens by foot, an hour to cross 18th-century Paris by carriage, an hour to cross industrial London by streetcar, and an hour to cross the contemporary *average metropolitan area* by car or rapid transit carries the cabalistic fascination that all reductionisms do, but it entirely misses the sociological point. The difference between the ancient Athenian pedestrian and the contemporary *pendulaire* resides precisely in the fact that by walking one can meet, and interact with, all sorts of people. Moreover, one can get some exercise that reduces cholesterol and the demand for complicated health systems to take care of circulatory disorders. Finally, one need not take note of train schedules, set up by employees of large-scale transport organizations that are part of huge public or private corporations that, in turn, have something to do with the overall economy and the class system of the city. Furthermore, one can escape the distinctive separation of workplace and abode that invokes a number of significant social and individual consequences.[6] If, after all this, I am told that the result always is that famous hour—a parameter, nonetheless, that I would like to see confirmed by more generalizable and empirically supported data than a couple of engineers' calculations—well, then I am not particularly impressed.

No doubt what has been called a *standard metropolitan area,* following a considerable number of studies that culminated during the late 1960s, actually is a new breed of urban animal. As Norman Gras once said, "The large town, the outstanding town . . . slowly grew into the economic metropolis" (Gras, 1922, p. 181). With astonishing insight, H. G. Wells wrote,

> The railway-begotten giant cities . . . in all probability [are] destined to such a process of dissection and diffusion as to amount almost to obliteration . . . within a measurable further space of years. These coming cities . . . will represent a new and entirely different phase of human distribution. . . . The city will diffuse itself until it has taken up considerable areas and many of the characteristics of what is now country. . . . The country will take itself many of the qualities of the city. The old antithesis . . . will cease, the boundary lines will altogether disappear.[7]

These reflections strongly suggest that it is important to think of the changes in contemporary urbanization as an evolutionary process, provided, of course, that we purge this term of its linear implications. During the 20th century, the urban form has undergone, and still is undergoing, profound changes. If we avoid the hyperbole of the apocalypses predicated one moment on explosions and the next on implosions, then we will recognize that the phases of this process trace the contour of an extraordinary evolutionary process in which the developments of one phase provide support for the next. In other words, this process has proven to be sustainable, although it is neither ideal nor costless. The trend toward increasing urbanism has accelerated during the century, and it shows no signs of slowing. Rather, indications are that powerful forces still are at work shaping our urban world in a consistent way, as suggested by Michael Cohen's "convergence theory" (Cohen, 1996, pp. 25-38). For these reasons, I speak of different "generations" of metropolises, one distinguishable from the other with simple analytical tools.[8]

City Users and the Second-Generation Metropolis

All things considered, *early metropolization* based on the emergence of the commuter population coexisted with traditional urban structures to a fairly large degree. However, the increased mobility of individuals, combined with higher income levels and greater leisure time, allowed the differentiation of a third population: *city users.* This population is composed of persons going to a city mainly to use its private and public services—shopping, movies, museums, restaurants, health and educational services. This is a swelling population that is having radical effects on the structure of cities and that actually uses localities in a rather uncontrolled way. There are cities that have very small populations of *inhabitants,* slightly larger populations of *commuters,* but vast populations of *users.* Venice is a typical case. It has a *resident* population (shrinking) of about 70,000, has a working population composed almost entirely of commuters, and on certain days is visited by a population of visitors as large as its resident population. The density of people becomes so high that Venice probably is the only city in the world to have pedestrian traffic lights. Venice is an extreme case, but many other cities of the world—and not only the so-called "cities of art"—experience this phenomenon. In fact, because their economies depend increasingly on these nonresident populations, most contemporary cities of all sizes want to attract their share of city users. Unlike the commuters, the users make use of—and sometimes *abuse*—the *public areas* of the city, more often than not in a rather barbaric way. Not surprisingly, at the end of 1989, the mayor of West Berlin declared that he was not worried about disposing of *Die Mauer* (the Berlin Wall) "because tourists will take it away."[9]

This phenomenon, however, is far from being limited to tourist cities or to Western countries. Cities such as Singapore have more visitors than inhabitants and are entirely geared to consumption. They are not different in many ways from New York, London, or Milano (where every year 300,000 Japanese visitors religiously visit the Last Supper as well as the fashion "golden triangle").

This consumer population is growing, but its actual size and composition are difficult to assess; our collective cognitive apparatus deals with a traditional city that already has undergone a profound mutation. The commuters have been around long enough for official statistics to track them, albeit not nearly as precisely as is done for inhabitants. City users are new, and the official statistics have not yet responded. Nonetheless, harsh competition among cities for hosting the Olympic Games, large exhibitions, and other world-class events provides hard evidence of the crucial importance attached to the city users population by local elites and the consequent growing importance of city marketing.

Sociologically, the population of users is difficult to define. An educated guess would assess it as being fairly differentiated, from suburban kids roaming and cruising on evenings and weekends; to middle class tourists and shoppers of all ages; to special groups such as soccer fans, concert goers, and exhibition attendees. A theater such as La Scala, traditionally the artistic and social temple of the Milanese population, is increasingly taken up, years in advance, by city users coming from faraway countries, a large portion of whom are co-citizens of Madame Butterfly (i.e., Japanese). Access to services is not limited to shopping or entertainment. Schools, universities, and educational establishments are a very important and growing part of the services provided by cities. Thus, the student population is becoming a sizable portion of the city users population, not only in traditional university towns such as Bologna, Cambridge, Pavia, Heidelberg, and Oxford but also in large cities with more differentiated economies. In Milano, a city that by traditional parameters has a population of 1.3 million, the university student population is close to 200,000, more than half of whom come regularly from out of town.

Although direct competition or conflict of users with inhabitants is not evident, *indirect competition* (in the sense in which the classical social ecology uses this term) is taking place. The user population is not attracted by residential areas except when the latter fall into the category of *picturesque,* but it heavily affects the spatial composition of central cities and of some specialized suburbs. Commercial and leisure areas of the city are most affected, with increasingly profound impacts on the global social structure of the city. Areas such as the Parisian *Quartier latin* and parts of Rome, London, New York, and scores of other cities teeming with discount stores, jean shops, fast food, and the omnipresent signs of the *rags multinational* tend to selectively filter out the original population of the neighborhoods, even when they initially constituted the local attractions in the first place. The same is happening in top commercial strips such as Rodeo Drive, Faubourg Saint Honoré, SoHo, and via Montenapoleone. But the population of city users is not limited to leisure or shopping. The city provides other services that can be used such as those connected with mass education or health. In a novel version of the "town-and-gown" medieval fights in cities with colleges and universities, students cause housing conflicts, particularly in neighborhoods (and on their fringes) called *college towns.* The type of metropolis that is growing out of the heightened gravitation of city users is the one we live in nowadays. It is very different from the city we are accustomed to dealing with in popular and scientific terms, and it can be defined as the *second-generation* (or *mature*) *metropolis.*

The Third-Generation Metropolis

Yet a fourth metropolitan population is forming. This is the very specialized population of *metropolitan businesspeople*—people who go to central cities to do business and establish professional contacts or businesspeople and professionals who visit their customers, conventioneers, consultants, and international managers. This fourth population is relatively small but growing. It is characterized by having access to considerable private and corporate money. It is a population of expert urbanites who tend to know their way around and who are very selective in terms of shopping and hotel and restaurant use as well as in the use of high-end cultural amenities such as concerts, exhibitions, and museums but also saunas and gyms. Increasingly, *business* and top-level *tourism* go together.

Both the city users and the metropolitan businesspeople are products of the service industry and of the globalization of the economy. A little-explored aspect of the service industry is the fact that whereas manufacturing industries move goods, services in large part require the shifting of persons. Despite a growing portion of services that can be delivered telematically, most services require face-to-face contacts, even when the partners are not terminal consumers, as in services to firms. Consulting, public relations, marketing, and the like all are activities that require intense and repeated face-to-face interaction.

The fourth population is increasingly constituting a transnational middle class living not in *a* city but rather *in cities* or *between cities.*[10] This affects the morphology and functions of all large cities well beyond the group's numerical weight. For the sake of classificatory completeness, we can call this still emerging metropolis the *third generation* (or *late*) *metropolis.*

The preceding analysis receives additional insights in the frame of Giddens's (1990) concept of "disembedding" as a trait constituent of what he calls *radical modernity,* a concept that I find more illuminating and analytically powerful than the current cult term of *postmodernity.*

Today's City of the Future

Class Structure in the New Metropolis

The positing of these four populations does not imply that more traditional class relations and conflicts have disappeared, although there is little doubt that they have undergone deep transformations undermining the classical socioecological factors of urban class relations. In purely numerical terms, of the four populations, the inhabitants tend to be by far the most disfavored by the overall dynamic. The commuters probably are shrinking or, at any rate, becoming a more composite type of social aggregate as even top-level coordination functions tend to move to the periphery of large conurbations and full-time employment decreases its share of total employment. All in all, the traditional class cleavages and solidarities, although still existing and perceptible, give way to new cleavages and group realignments.

But what is the relation of the four populations to a fifth one, the so-called *immigrants*—which, of course, usually means *poor immigrants*? The glittering image of the city of consumers seems at odds with the growth of visibly impoverished populations in most cities of the world. The easy answer is that all four populations are stratified, albeit at a slanted angle. In general, the immigrant population is attracted by the possibility of finding jobs or making livings. Thus, it will be in a larger proportion among the inhabitants and least of all or nonexistent among the metropolitan businesspeople.

A more complex answer also exists. The metropolitan transformation is such that the entire class structure is changing. Let us consider the class structure of the industrial city in which new social classes were created. With the rise of industrialism, cities became ugly places of toil, human misery, and social and physical ills—mean streets as well as mean air (*mala aria urbana* in the original meaning of malaria). The average height of the population went down from the 5 feet 9 inches of the late agricultural European (1,350-1,150 years ago) to the 5 feet 7 inches of the early industrial European of 150 years ago ("Height and Technological Change," 1994). Mortality was so high that in England the poor were buried in layers one on top of the other in the same pit. A new societal division of labor was being hammered out, and it produced two distinctive classes that did not exist before. Around the factory, the urban proletariat settled in working class neighborhoods in which labor organizations and parties transferred horizontally the class ties and solidarities established in the factory industry. Plant, neighborhood, and political organizations constituted the strategic niche of the working class city, a city existing in conflicting symbiosis with the bourgeois city contemporarily created by the class of industrial capitalists. Although they never became all encompassing as the Marxian theory had assumed, the two industrial classes shaped the physical and social environments of the industrial city.

We know a lot about the social stratification of the inhabitants of the industrial city, and this knowledge carries over to the residential portion of our contemporary metropolis. We know a lot less about the social traits of the commuter population that grew during the 20th century, but we know a great deal, especially from the socioecological point of view, about the social morphology of human settlements in metropolitan areas or first-generation metropolises. For example, we can surmise that the commuter population in the United States tends to be middle class and is different from the one in Europe, where it tends to be lower class.

Certainly, it is difficult to categorize the social stratification of these new populations. This alerts us to the fact that the discourse on the class structure of the new city is increasingly complex. The difficulties stem largely from the fact that to understand the new social stratification of cities, we have to scale up our level of observation. In other words, the new stratification cannot be understood at the individual city level only. What has been observed at the individual city level is a growing polarization, but the origins of this polarization are mainly exogenous. The services required by city users and metropolitan businesspeople are largely manned by marginal workers. Very often, these workers come from the ranks of the newcomers. Saskia Sassen shows very clearly that a relation exists between the economy of the global city and the "vast supply of low-wage jobs required by high-income gentrification in both its residential and commercial sectors" (Sassen, 1991, p. 9).

There are cities in which this process appears in an extreme and, therefore, pure form. One of these is Singapore, a city living almost entirely on the services offered to a transient population in which most of the lower level jobs are held by mainland Malaysians and other immigrant populations (Bonazzi, 1996). In other cities, hampered by national hinterlands and by the more complex maze of historical stratification, this dialectic is less evident, but it is becoming increasingly clear. Still, it would be preposterous to extend the argument to the point of seeing in the contrast between city users and inhabitants a new class conflict somehow reproducing the traditional one between the factory owners and the factory workers. No doubt, the incipient class polarization noted in large urban areas is strongly connected to the impact of the new populations of metropolitan users as opposed to city *dwellers* or *workers*. Dwellers or workers, especially in the lower income brackets, are more ecologically bound. The competition among these several populations favors the more mobile ones; the urban inhabitants tend to be at a disadvantage.

The most crucial aspect of the new urban social morphology is the creation of a new class relatively free of location. Very little is known about this new class except that it leaves increasingly significant traces. The growth of the fourth population, metropolitan businesspeople, signals another very important phenomenon, namely the internationalization or globalization of metropolitan centers. The fourth population is increasingly constituting a new highly mobile middle class, and it affects the morphology and functions of all large urban centers. This is a fairly heterogeneous social class and is increasingly identifiable—managers[11] of multiloci enterprises, both private *and* public, such as the large number of *international organizations* (e.g., United Nations; International Labor Organization; United Nations Educational, Scientific, and Cultural Organization; Organization for Economic Cooperation and Development; Food and Agriculture Organization; World Health Organization; the growing family of European governmental bodies), businesspeople, international consultants, academics, performers, and sportspeople. This population requires fairly similar services all over the world—hotels, office and meeting places, restaurants, shopping centers.

Among the postcards that the traveler can buy at any airport newsstand the world over is one showing the local *skyline*. Increasingly, these skylines, as well as the urban areas they depict, tend to look alike. This is not surprising; these areas are not the product of *national economies* but rather are a *segmental* unit of a larger entity (Friedmann & Wolff, 1982). Hotels, offices, and commercial centers built by the same companies in many cities go together with the standardization of local shops catering to an increasingly homogeneous transnational population of urban travelers.

In Europe, this trend has been slowed by the strength of national urban cultures. For centuries, the top-ranking cities of European urban systems embodied the specificity of local culture and traditions. European national and regional capitals of the 19th century symbolized the climax of this dynamic; Vienna, Paris, London, Berlin, Milano, and

Florence offered themselves to the learned traveler as distinct worlds with languages, architecture, cultural institutions, and social mores proudly displaying the best of their respective national or regional characters. The facade of this identity still is standing, but homogenization is at work. Prince Charles's battle against the "ugliness" of the new urban architecture in England is a telling example. The London skyline, as well as those of other European cities, displays vividly the superimposition of the old and new architectural patterns ("Whose Britain Is It?", 1989). Resistance to fast food shops in several European capitals (Rome and Paris in particular), far from being insignificant, is a sharp indicator of the conflict between traditional national identity, which includes the culinary culture, and aggressive modern multinationals whose commercial success is based precisely on extremely fastidious imposition of product and labor force standardization.

Who Are the Stakeholders in the New Metropolis?

Another consequence is that the type and level of municipal welfare services organized under the assumption that conflicting classes had to live together in the same area increasingly lose legitimation. In the traditional city built around the two major classes of the Fordist system, the ruling class had to strike an institutional compromise by giving back through taxation. At the same time, the internal sociological borders of the city were more clearly defined as part of this compromise, whereas the factory provided another set of social boundaries, both inside and in relation to the city. Today, no mayor wants to have the users visiting his or her city be bothered by the homeless. Few mayors of contemporary cities, however, will try to solve this problem with increasing social welfare. The preferred solution is the one patented by Mayor Rudolph Giuliani of New York City; that is, keep the homeless out of the crucial areas and sweep them to the periphery. Thus, the issue of class structure relates very strictly to the politics of the new metropolis.

In the competition among these several populations and related urban functions, the residential function and the urban inhabitants tend to be on the losing side. The entire philosophy of local government, however, is based on various degrees of self-government by city dwellers. If the resident population is less and less relevant from the numerical and economic points of view, then one serious and far-reaching consequence—already behind many manifestations of the urban crisis—is a de facto disenfranchising of the urban dweller. Local governments are elected by residents, but the economic interests of the metropolis are increasingly dependent on populations not politically accountable from the point of view of the city itself.

The issue has to be raised: What is the basis of citizenship in the new metropolis? There is a considerable discussion of *social exclusion* of the poorer strata of the population including the more marginalized groups of newcomers. The inhabitants of a city, however, might well suffer a particular form of *exclusion* as their territory is invaded by the new transient populations of consumers.

Traditionally, municipal citizenship has been based on some residential attribute automatically connected with access to public places. Most of local governmental practices and institutions have been built on this assumption (backed by concrete and consolidated historical experience) that inhabiting a place, and particularly owning real estate there, creates a strong propensity to make the necessary private contribution to common welfare. To use a contemporary formula, residence allows one *to have a stake in the community.* Moreover, the physical model taken as a benchmark for democracy in the city includes a central open space such as the agora in the classical polis or the piazza in the medieval Italian commune, where the citizen assembly can meet to discuss, exercise jurisdiction, and make decisions. Thus, in the historical lore (popular as well as scholarly), the physical shape of this place became synonymous with democracy. Its disappearance or transformation has become synonymous with the waning of democracy and a recurrent theme of democratic nostalgia fed by the undeniable oligarchic trends implicit in mass democracy.

Non-Lieu Urban Realms

The new urban world is invaded by an alien monster, not the familiar BEM (bug-eyed monster) but rather the WEM (West Edmonton Mall). The WEM is only one of a new breed of urban features that cater to the city users population and it is transforming the urban landscape. In Margaret Crawford's words, malls are the product of the "science of malling," a new applied discipline created by real estate promoters with the help of architects and lawyers (Crawford, 1992).

Mall, short for *shopping mall,* is a word that derives from the Italian Renaissance game *pallamaglio,* played in streets where shops and amusements were located, such as London's *Pall Mall.* The name stuck for an integrated real estate complex with shops, restaurants, and other services when the diffusion of cars pushed for functional differentiation in the city and made access to mixed areas (e.g., the old city centers) increasingly difficult. *The mall* has become a central feature of the city of users. Whereas in the past malls were built around cities, now cities grow around malls. Besides, the mall epitomizes the new category of "non-places."[12] As happens with catchwords, the obsessive repetition of the non-place idea popularized by Marc Auge reveals its inherent analytical weakness (Auge, 1995). We instinctively understand what is meant by non-places—subway stations, airports, supermarkets, malls. However, we understand it so well not because they are non-places but rather because they are *the* places of the city we live in today. Non-places are nothing less than the typical places of the city of our times. True, they are abstract, impersonal, anonymous, and perhaps anomic, but they are this way because we want them to be this way. Our own society, and in particular the type of people who write and read books about non-places, has invested a staggering amount of economic, technological, and institutional resources to create these "places non-places," where an individual can with ease be a member of the lonely crowd.

Normally in the debate about the public sphere of contemporary societies, these places non-places are opposed to previous types of public places—*la piazza* in the European tradition or *main street* in the American one. This comparison, and the cultural and social consequences inferred from the differences, is ideological and prospective rather than real. Can we be sure that in the transformation of the late Roman city, the diffusion of basilicas did not bring about similar anxieties (Gombrich, 1966)? Where had all the hustle and bustle of classical temples gone? Where were the lively crowds of donors, peddlers, and officiants? What happened to the animal whose ceremonial sacrifices diffused a constant barbecue-type smell to titillate the Olympic gods with the perfume of burning meat? One of the tasks of the *vestales* in the Roman temple was the *stercoratio,* which was to clean from the premises the manure of the cows pasturing around and in the temples.

Particularly dehumanizing to Auge and similar researchers of the times must have been the disappearance of statues from sacred places due to the religious belief and communication strategy of the Christian church, in the end exploded in the iconoclastic movement to distinguish its basilicas from the Roman temples (Gombrich, 1966). Yet today, we would hardly dare to classify the early Christian churches as non-places. Thus, I am convinced that a good part of the negative connotations of non-places is due to a specious characterization that, as happens to all generations, we have not yet come to consider as ours. Contrary to the relative and misleading stability of the natural environment from one generation to the next, the urban scene is changeable, particularly so during historical times such as the one in which we live. Rather than being passed from one generation to the next, the built environment changes continuously and appears alien to its older inhabitants.

Democrazia di Piazza

More or less in the same vein, the new non-places, particularly the commercial malls, have lost the public character of la piazza. Access is not universal, and the function of the agora or place of political debate, which was typical of public places of the past, is not present. One obvious response is to wait and see. More analytically, the function of la piazza, which Carlo Cattaneo described as "the heart of cities" (Cattaneo, 1972), has been overemphasized. There are many indications that the piazza was not so public and democratic as we often imply. In the cradle of civic democracy, 5th-century Athens, crowds could be anonymous to the point that not even slaves could be easily recognized. Recent Greek scholarship has challenged the concept of a strictly communitarian interpretation of Athenian society. Edward Cohen's highly significant work points to a number of historical facts. Even in ancient Athens, face-to-face relations were not as frequent and diffused as a Norman Rockwell version of the ancient town would depict them:

> For ancient historians, Athens is the incarnation of "the Athenian citizen body" . . . , [and] they have appropriated Laslett's conceptualization of English village life to identify Athens as a "model of face-to-face society" where the entire population knew one another intimately and interacted closely on a society-wide basis. . . . This false premise has infected a multitude of significant contemporary Western scholars and thinkers—from communitarians to Kuhnians. (Cohen, 1998, p. 10)

Actually, the interrelation between private and public is more complex than it appears. The public space not only can be an extension or a complement of the private one but also can be its dialectical opposite. The final result can be obtained by subtraction when the public space is considered as residual to one's sphere of life. "Cities with no civic communities" is a concept discussed vividly by Yanis, particularly with reference to the southern European town.[13] I was once shocked by a fairly rich friend who lives in a southern Italian town of middle size. His house is more than simply beautiful; it is a stately ancient mansion with an old garden with century-old trees and ancient sophisticated internal decor, complemented by extremely comfortable modern interiors and appliances. The only really repulsive part is the external wall, a squalid, dirty, cheap-looking construction not different from those that could be seen in any slum area of the same city. When I inquired about the contradiction, my friend very serenely answered, "I do not look at my house from the outside." He was not in the least interested in the "public eye" and its conscience. Quite the contrary, he implied, any embellishment directed to outsiders would be a waste of money. The lack of a public eye or the generalized exclusion probably is the heritage of an ancient mode of conceiving the polis, one in which public space does not exist because the city was a sum of private abodes set apart by family religions (Fustel de Coulanges, 1924)—the pre-politicization of individuals (Bairoch, 1985; Glotz, 1926).

Does It Take a Village?

This attitude is not necessarily linked to the concept of "secure" public spaces. Most southern European towns are not as dangerous as some of the dense downtowns, or even suburban areas, of large northern metropolises. They do possess "*the civilized quality of an urban area*" consisting in the "freedom of walking about and looking around" (Ciuffini, 1994, p. 6, emphasis in original), which nobody will challenge in Naples or Athens, no matter how dirty the streets. Actually, one can face hostile environments if embellished with armor such as the "Home Moat," Tom Wolfe's "favorite piece of landscaping of the happy worker suburbs of the American West." Wolfe (1968) says,

> It is about [3] feet wide and a foot and [a] half deep. Instructions for placing rocks, flowers, and shrubs are available. The Home Moat is a psychological safeguard against the intrusion of the outside world. The Home Moat guards against the fear that *It* is going to creep up in the night and press its nose against your picture window. (p. 7)

Although the quality of life appears to be very high in American suburbs, the "freedom of walking about and looking around" should not be taken for granted. Depending on who one is and how one dresses, or simply because one walks (jogging with the appropriate gear is all right, as is driving), one might end up frisked, if not actually shot at.

The lack of a public life in southern European urbanized areas has been pointed out by authors such as Percy Allum in his book on Naples (Allum, 1973) and Robert Putnam in his book on Italian regional policies (Putnam, 1993), not to mention Edward Banfield's pathbreaking (if controversial) research on the moral basis of backward societies (Banfield, 1958). The *long durée,* rather than contingent situations, would explain the current lack of a public eye in large parts of the urban world. Of course, the long durée means reinforcing mechanisms such as the different relations between the merchant emerging classes and the feudality in southern and northern Europe. Alessandro Pizzorno remarks that in Italy, where the "nobles live in the city, there appears a phenomenon of emulation in consumption . . . which leads people to spend more than in those countries where the nobility remains to a large extent in the country" (Pizzorno, 1973, p. 123).

The issue is not new and goes back, at least, to Alexis de Tocqueville's work:

> There are countries in Europe where the native considers himself as a kind of settler, indifferent to the fate of the spot which he inhabits. The greatest changes are effected there without his concurrence and (unless chance may have apprised him of the event) without his knowledge; nay, more, the condition of his village, the police of his street, [and] the repairs of the church or the parsonage do not concern him, for he looks upon all these things as unconnected with himself and as the property of a powerful stranger whom he calls the government. (de Tocqueville, 1961, p. 6)

The Price of *Pax Municipalis*

A great deal of urban literature grapples with the problem of severing the physical and social levels after having mixed them irrevocably. The impression one receives is that this literature is dominated by a fairly romantic and not always realistic view of public squares. The lore about agora or la piazza derives, not surprisingly, from the medieval commune or from the idea we have of its grassroots democracy. Nobody can deny the reality of public assemblies in the communal *brolo* or market; its democratic and universal character has been vastly romanticized. As Henri Pirenne observes, medieval communes were fairly nondemocratic, with stratification fostered by the complex corporatist system

(Pirenne, 1980). Marino Berengo's studies on Lucca and the *città di antico regime* stress this point by underlining the working class role of apprentices (Berengo, 1965; see also Pirenne, 1980; Romano, 1993; Rugge, 1992). The most important testimony comes from Nicolò Machiavelli himself, with his notable distinction between la piazza (where the shallow political expressions take place) and *il palazzo* (where real political decisions are made). As Marco Romano (1993) notes,

> The spatial theme of "la piazza" has so conspicuous antecedents—the Greek agora, the Roman forum, mosques' courtyards, and Christian cloisters—to suggest that it is a constant archetype on cities' horizon. In Europe, this theme had disappeared as early as the 6th Century CE. The square theme should not be taken for granted. Quite the contrary, it is the result of an effort, a creative tension of several generations which in Europe will take hold very slowly and with difficulty. The interiorization of the interurban market in the city is by no means an obvious solution. In the Islamic city, it will always remain on the outskirts, separated from the internal *bazaar*. (pp. 79-80, emphasis in original)

Cattaneo (1972), in writing that the square is not "an empty space . . . but [rather] a pleasant and useful place for citizenship" (p. 69), had in mind a Simmelian density of contacts. He opposed the construction of a large square because an "unmoderate space, one which separates excessively the built parts, would interpose a continuous and inevitable obstacle to all social relations, increasing the opposite nuisance of seasons, rain, sun, winds, and fog" (p. 69). This unconventional idea coincides strikingly with Maurice Culot's stern criticism of functional conceptions of the city. "La separation des fonctions," writes Culot (1994) advocating a holistic principle, "est l'antithèse de la ville" (p. 11).

The functional element is very important. If in our time public spaces tend to be reduced as a consequence of real estate competition for valuable lots, the symmetrical physical situation of huge empty spaces does not necessarily mean that public life is encouraged. Ceausescu's Bucuresti, like many other capitals of the "Socialist city," had a bonanza of huge empty squares that very effectively discouraged the temptation to cross them, let alone to linger about in a flaneur mood. The same applies to the sad dusty parks of new housing projects the world over.

The idea of the square as a very intimate and *gemütlich* place protected by the natural elements is indeed typical of the medieval European city. The current popular image of the city's *d'antico regime* fails to acknowledge the high degree of violence and actual armed conflict experienced daily by citizens. We all love the rugged spiked look of Volterra with its many towers. Few know that in Volterra (as in hundreds of other towns where the towers have been partially or totally destroyed), these medieval buildings functioned to protect the inhabitants from their neighbors' similar fortifications a few yards down the road. We sometimes complain about strains in neighbor relations, but in medieval Volterra, this often meant trading arrows and other deadly projectiles. Medieval communes were, in general, very factious societies, and the much vaunted fact that their citizens were armed certainly was not conducive to dialogue in public spaces, as Romeo and Juliet found out. Another masterpiece of world literature, Dante's *La Divina Commedia,* originated from the then normal fact that members of the losing factions (or of the opposition) were either killed or exiled for life, as happened to Alighieri.

I am sure that this will be considered unbearably heretic, but there is nothing in Cattaneo's description of the square that would not apply to a well-designed *mall.* I know what enraged critics will say. The mall is a built environment full of deceptions, and its public aspect is only apparent. Malls, as well as other similar non-places (e.g., exhibition centers, airports, hotels), are full of architectural lures and deceptions. With very similar intent, architects of the past built churches, squares, and entire cities. The difference is that we take for granted the architectural deceptions of the past. Sometimes they elude us. We feel the grandeur, but we miss the subtleties. In the contemporary, commodified, and thematic urban scene built for our consumption, we are continuously aware of the ongoing deception. The main deception, it is said, is the apparent public character of these new places that are in fact private compounds, well guarded by nosy (if discreet) security systems. Two uniformed *carabinieri* have been a constant diurnal presence in Piazza del Duomo in Milano, as in other major Italian squares, reminiscent of the medieval *armigeri* that helped guarantee the tough *pax* of the medieval town.[14] In a sense, it is precisely the *lack* of police control that has taken away the public character from the piazzas of many European cities and turned them into more impersonal and threatening places for a transient population including the marginalized immigrants.

No doubt, the new city needs a redefinition of public spaces. The mall has become the hated counterpart of the public and "political" square; digital technology has become the new vehicle to dehumanization.[15] As in many other similar cases of *laudationes temporis actis,* we have to resist strongly the tendency to idealize the public function of la piazza along all of history. During limited periods of time, and in a restricted number of societies with well-developed *burgherlich* characters, the square or other similar public place has been the "heart of the city." As Romano points out, however, the public character of these places, as spaces for democracy, had to be negotiated. Moreover, this is not the only space in which democracy works. In some societies, la piazza was the main public arena, whereas today there are other spaces, including the immaterial ones, in which the democracy of our times has to be renegotiated. In a city where a growing part of the population is transient and largely service oriented, new forms of participation and, even before that, of identification will have to be worked out. As already has been hinted, the change is so deep that it affects the very heart of our understanding of the city, namely our cognitive tools.

■ Consequences for the Urban Sciences

In all respectable scientific enterprises, including the sciences of the city (i.e., the various urbanological enterprises from city history to city planning), knowledge is based on the quality of observational tools. Much can be said in favor of direct observation of city life; the city is an eminently visible phenomenon, although literary and occasional onlookers tend to be greatly misled by what they can see before them. A large part of social life takes place out of view. Valuable as it might be, direct observation tells only a partial story about urban society. After all, society is literally invisible and can only be inferred by the traces it leaves.

The bulk of our knowledge of the social or invisible aspects of the city comes from the large body of systematic data collection that we call *statistics.* In all countries, the majority of statistics pertaining to cities are composed of census-type data. These data are universally used by planners, local administrators, media professionals, and scholars, all of whom consider them to be endowed with a high degree of validity. Undoubtedly, they appear as data of a "harder" nature than the "soft" sample survey data. However, census-type data are not observational data at all but rather recordings of individual verbal behavior, with all the limitations that affect survey results and a few additional ones as well. The quality of the individual data collection is not very high. The advantage to the researcher is that census-type data are, by definition, exhaustive of a given population and, therefore, can be used to construct ecologically based maps or tables.

Better coverage and, in some cases, even more reliable data come from *process-produced data* such as the various population registers. These are the traces that the human particle leaves during its passage through the bureaucratic maze. As I have indicated elsewhere, and as every urban scholar knows very well, these traces, which are part of the organizational knowledge of our times, often are extremely useful and reliable.

Both primary data for urban research (e.g., census-derived data) and process-produced data, in addition to their well-known technical limitations, have a further weakness deriving from the changes in urban social structure outlined here. Census data and process-produced data based on administrative records are collected at the place in which people live. Census schedules are very specific about this; their instructions call for the identification of the persons who are in a given location "at midnight" of the census tallying day. This means that all these data provide an image of the *dormant city.* Thus, the immense variety of thematic maps of social aspects of the city, which scholars have been using to draw their social models of cities ever since the Chicago school and even earlier (starting at least from John Graunt and Sir William Petty during the 17th century[16]), produce snapshots of the urban population *asleep.*

In the past, the *sleeping* city and the *active* one overlapped to a very large degree. To be more precise, one could say that the movements of the urban human molecule were of a sufficiently short wavelength to stay within the power of resolution required by the observer. The map of the sleeping city would not differ considerably from the one of the working city if the unit of observation was the block or the neighborhood. With the rise of the first-generation metropolis, and increasingly with the growth of the city users population, the difference between the two becomes progressively more significant. Although the nonresident populations become more and more visible on the streets, they are invisible in the statistics. Of course, these populations are tallied, in one way or another, through directories, airport, hotel, and exhibition checkpoints and rosters, credit reports, and a number of other ways. There is no systematic or public way, however, in which these new populations are measured or even estimated exhaustively. Quite apart from the fact that most of these data are patently beyond the reach of social scientists (who, by the way, would be much more careful and innocuous users of data of this type than are the snoopers who currently get their hands on them), what is really damaging for the knowledge of urban phenomena is the quantitative disparity between these and the more traditional data. It is fair to say that inhabitants account for the great majority of the available data, commuters for a very small proportion, and the remaining populations for very little indeed. Data about the users population sometimes (but not very often) might be of high quality (e.g., top-level survey data on travelers and on hotel and exhibition guests), but they are *scattered* data. This means that we are looking at the city with biased eyes, and this is no minor problem.

In a sense, we are looking at the sky with only visible light wavelength telescopes. This is not to say that the things we would see are not there, but the sidereal space—as well as the city—is full of events that escape our observation if we limit ourselves to instruments capable of producing measurements within only a restricted band of vision. Astronomers and physicists have given ample proof that we would be utterly misled if we were to base our knowledge of the physical universe exclusively on the images of visible objects. As paradoxical as it might sound, an equally great mistake will befall us if we persevere in basing our knowledge of the contemporary city only on the analysis of its inhabitants.

NOTES

1. The boosting of city images, for both political and commercial reasons, can be traced to ancient cities, perhaps to the very origin of the city; the urban dweller always has felt his or her superiority over the rest of the world. However, the unabashed commodification of cities as sales objects has become a matter of course only in very recent years and can be easily dated to the early 1990s (Ashworth & Voogd, 1990; van den Berg et al., 1990; Falk, 1989; Rand McNally, 1989). Karl Marx certainly would have quoted this as an extreme case of "exploitation not veiled by religious and political illusions, naked, shameless, direct, brutal exploitation" (Marx & Engels, 1932, p. 324).

2. The list is long, but not as long as one would expect. Only recently have social scientists delineated the contours of the new urbanization in a way not hampered by mental categories created on disused urban forms, namely works making a deliberate effort to look at the emerging urban form by connecting changes in the economy with

technological innovation and social change. Among the most important ones, in addition to the initial work of Friedmann and Wolff (Friedmann, 1986; Friedmann & Wolff, 1982), I would list Bagnasco (1986), Bianchini and Parkinson (1993), Castells (1996), Castells and Hall (1994), Logan and Molotch (1987), Masser et al. (1992), Pumain and Godard (1996), Sassen (1991, 1994, 1995), Sorkin (1994), and Sudjic (1993).

3. Castells (1972) was the first to criticize this definition as well as the analogous one by Childe (1950). In his critique, Castells (1972) stresses that concentration is not equivalent to urbanization but rather one aspect of a larger social phenomenon that we can call urbanization. More precisely, one could say that concentration is the observable aspect of social transformation that for a long historical period has coincided with the formation and development of cities, that is, transformations having to do with division of labor and social relations of production (pp. 35 ff.).

4. The long debate on city and country in the Marxist tradition has found a topical moment in the heated *querelle* between Lefebvre and Castells (Castells, 1972; Lefebvre, 1968, 1970, 1972, 1973, 1974). For a useful synthesis of the debate, see also Saunders (1986, pp. 152-182). In fact, it is important to distinguish between Western Marxism, generally favorable to urban life, from that of the Third World, exemplified in its most extreme and dramatic form by the urban luddism of Pol Pot. In the orthodoxy of European Marxism, the contradictions between country and city can be overcome by an extension to the country of urban traits, an attitude relevant to the serious difficulties of agriculture in many Socialist countries and in the Soviet Union in particular (see, among others, Popkin, 1979). Although less generally known than the reverse Pol-Potian drama, a chilling example of this general orientation comes from the campaign of destruction of the beautiful traditional villages of Romania organized, but fortunately only partially implemented, by the Ceausescu regime. The plan was to transfer all of the rural population to multistoried "blockhouses," many of which were then constructed without sanitation. See Grigorescu (1990).

5. I will not even try to make a list of these announcements, an exegetical undertaking at the periphery of the main argument of this work. See the interesting comment on this theme in Harvey's (1989, pp. 4-9) introduction to his work on postmodernity.

6. *Boulot, metro, dodo* is the Parisian rigmarole to describe the drudgery of the commuter's life, spent entirely going from work to travel to sleep.

7. Berry (1976, p. 24) deserves an applause for having recalled this extraordinary statement by Wells (1905/1962).

8. The classificatory scheme I propose is based on three simple dichotomous variables that can be measured on individuals, thereby identifying four populations. By *population,* I mean an aggregate of individuals for which there is no need to posit any strong assumption of collective rationality.

	Live	*Work*	*Consume*
A. Inhabitants	Yes	Yes/No	(Yes)
B. Commuters	No	Yes	(Yes)
C. City users	No	No	Yes
D. Metropolitan businesspersons	No	Yes	Yes

The emergence of the B, C, and D populations identifies three successive metropolitan generations. The signs of parenthesis and slash identify fuzzy cells.

9. East Berliners seem to have been more organized; they apparently created a corporation to sell chunks of Die Mauer. Whether by sale or by theft, the result is the same; a dramatically important piece of the *built environment* has been *used.* One is only too easily reminded of the prophecy by Marx and Engels (1932) that all Chinese walls would be "battered by the heavy artillery of the light prices of commodities" (p. 38). Not all the predictions of these authors worked out equally well.

10. Thanks go to the late Roy Drewett for this formulation. It was easy for him as an outstanding member of this new population.

11. There has been very little research conducted on this population, at least to my knowledge, and even less so by urban sociologists. But see an interesting book by Marceau (1989).

12. The idea was originally put forward by Webber with the phrase "non-place urban realm," but it has been diffused by Augé to the point that the two names are associated (Augé, 1995; Webber, 1964).

13. "D'aucuns diront que dans les villes méridionales, ce n'est pas l'exclusion sociale de certains groupes qui est préoccupante, mais une exclusion générale" ("Les Carrefours de la Science," 1994, p. 173).

14. Pirenne (1980) is quite firm about the Mayor Rudolph Giuliani type of social control and security in the city of merchants. On the other hand, secure cities, particularly at night but also during the day, as a premise to civilized urban political life are a very recent occurrence. Until very recent times, the abusively quoted *flâneur* would have risked his purse, if not his neck, in most historical urban scenes.

15. This is but a contemporary chapter in a long story that began with the ancient Greeks' fear of Prometheus and its *tekné* as a betrayal of the state of nature of the Golden Age (Vegetti, 1976, pp. 12 ff.).

16. In fact, Graunt (1662/1975, chap. 8) was concerned about the possible distortions introduced by immigration dynamics in his analysis (see also Lazarsfeld, 1961)

REFERENCES

Allum, P. (1973). *Politics and society in post-war Naples.* Cambridge, UK: Cambridge University Press.

Ashworth, G. J., & Voogd, H. (1990). *Selling the city: Marketing approaches in public sector urban planning.* New York: Belhaven.

Augé, M. (1995). *Non-places: Introduction to an anthropology of supermodernity* (English trans.). London: Verso.

Bagnasco, A. (1986). *Torino: Un profilo sociologico.* Turin, Italy: Einaudi.

Bairoch, P. (1985). *De Jericho à Mexico: Villes et économie dans l'histoire.* Paris: Editions Gallimard.

Banfield, E. (1958). *The moral basis of a backward society.* Glencoe, IL: Free Press.

Berengo, M. (1965). *Nobili e mercanti nella Lucca del '500.* Turin, Italy: Einaudi.

Berry, B. (Ed.). (1976). *Urbanization and counter-urbanization* (Urban Affairs Annual Review, No. 11). Beverly Hills, CA: Sage.

Bianchini, F., & Parkinson, M. (Eds.). (1993). *Cultural policy and urban regeneration: The West European experience.* Manchester, UK: Manchester University Press.

Bonazzi, G. (1996). *Lettera da Singapore ovvero il Terzo Capitalismo.* Bologna, Italy: Il Mulino.

Castells, M. (1972). *La question urbaine.* Paris: Maspero.

Castells, M. (1996). *The information age: Economy, society, and culture.* Oxford, UK: Blackwell.

Castells, M., & Hall, P. (1994). *Technopoles of the world.* London: Routledge.

Cattaneo, C. (1972). *La cittá come principio.* Venice, Italy: Marsilio.

Childe, G. C. (1950). The urban revolution. *Town Planning Review, 21,* 17-32.

Ciuffini, C. M. (1994, September). Paper presented at the "Urban Innovations and Medium-Sized Cities" conference, Oviedo, Spain.

Cohen, E. E. (1998). A modern myth: Classical Athens as a "face-to-face" society. *Common Knowledge, 5*(3), 11.

Cohen, M. A. (1996). The hypothesis of urban convergence: Are cities in the North and South becoming more alike in an age of globalization? In M. A. Cohen, B. Aruble, J. S. Tulchin, & A. M. Garland. (Eds.), *Preparing for the urban future: Global pressures and local forces* (pp. 25-38). Baltimore, MD: Johns Hopkins University Press.

Crawford, M. (1992). The world in a shopping mall. In M. Sorkin, *Variations on a theme park: The new American city and the end of public space* (pp. 3-30). New York: Noonday.

Culot, M. (1994, September). *Désir et désirabilité de la ville.* Paper presented at the "Urban Innovations and Medium-Sized Cities" conference, Oviedo, Spain.

de Tocqueville, A. (1961). *Democracy in America.* New York: Vintage.

Durkheim, E. (1898). *Les règles de la méthode sociologique.* Paris: Presses Universitaires de France.

Falk, S. (1989). Marketing your city's service advantage. *Western City, 65*(6).

Fortune Magazine Editors. (1958). *The exploding metropolis.* Garden City, NY: Doubleday.

Friedmann, J. (1986). The world city hypothesis. *Development and Change, 17,* 69-84.

Friedmann, J., & Wolff, G. (1982). World city formation: An agenda for research and action. *International Journal of Urban and Regional Research, 6,* 309-344.

Fustel de Coulanges, N. D. (1924). *La città antica.* Florence, Italy: Vallecchi.

Giddens, A. (1990). *The consequences of modernity.* Stanford, CA: Stanford University Press.

Glotz, G. (1926). *La cité grecque.* Paris: L'évolution de L'humanité.

Gombrich, E. H. (1966). *The story of art.* New York: Oxford University Press.

Gottman, J. (1961). *Megalopolis.* Cambridge, MA: MIT Press.

Gras, N. S. B. (1922). *An introduction to economic history.* New York: Harper.

Graunt, J. (1975). *Natural and political observations mentioned in a following index and made upon the bills of mortality.* New York: Arno Press. (Originally published in 1662)

Grigorescu, P. (1990, February 28). *Città e campagna nella Romania contemporanea.* Seminar, Dipartimento di Sociologia della Universitá degli studi di Milano.

Harvey, D. (1989). *The condition of postmodernity.* Oxford, UK: Basil Blackwell.

Height and technological change. (1994, October). *Scientific American,* pp. 96-97.

Hicks, U. (1974). *The large city: A world problem.* London: Macmillan.

Lazarsfeld, P. F. (1961). Notes on the history of quantification in sociology: Trends, sources, and problems. *Isis, 52,* 277-333.

Lefebvre, H. (1968). *Le droit à la ville.* Paris: Anthropos.

Lefebvre, H. (1970). *Du rural et de l'urbain.* Paris: Anthropos.

Lefebvre, H. (1972). *La révolution urbaine.* Paris: Gallimard.

Lefebvre, H. (1973). *La pensée marxiste et la ville.* Paris: Castermann.

Lefebvre, H. (1974). *La production de l'espace.* Paris: Anthropos.

Les carrefours de la science et de la culture. (1994). In *En quête d'Europe* (Introduction de Jacques Delors). Rennes, France: Apogee.

Logan, J. R., & Molotch, H. (1987). *Urban fortunes.* Berkeley: University of California Press.

Marceau, J. (1989). *A family business? The making of an international business elite.* Cambridge, UK: Cambridge University Press.

Martinotti, G. (1996). Four populations: Human settlements and social morphology in contemporary metropolis. *European Review, 4*(1), 3-23.

Martinotti, G. (1997a). *Metropoli: La nuova morfologia sociale della città.* Bologna, Italy: Il Mulino. (Originally published 1992)

Martinotti, G. (1997b). *Perceiving, conceiving, achieving the sustainable city: A synthesis report.* Dublin, Ireland: Loughlingstown.

Marx, K., & Engels, F. (1932). The communist manifesto. In K. Marx, *Capital and other writings* (with an introduction by M. Eastman, Ed.) (pp. 315-355). New York: Modern Library.

Masser, I., et al. (1992). *The geography of Europe's futures.* London: Belhaven.

Meyer-Eckhardt, V. (1946). *I mobili del signor Berthélemy* (Italian trans.). Milano, Italy: Rizzoli.

Pirenne, H. (1980). *Le città del medioevo.* Bari, Italy: Laterza.

Pizzorno, A. (1973). Three types of urban social structure and the development of industrial society. In G. Germani (Ed.), *Modernization, urbanization, and the urban crisis.* Boston: Little, Brown.

Popkin, S. L. (1979). *The rational peasant: Political economy of rural society in Vietnam.* Berkeley: University of California Press.

Pumain, D., & Godard, F. (1996). *Données urbaines.* Paris: Anthropos.

Putnam, R. D. (1993). *Making democracy work: Civic traditions in modern Italy.* Princeton, NJ: Princeton University Press.

Rand McNally. (1989). *Rand McNally sales and marketing city/county planning atlas.* Skokie, IL: Author.

Romano, M. (1993). *L'estetica della città europea: Forme e immagini.* Turin, Italy: Einaudi.

Rugge, F. (Ed.). (1992). *I regimi della città.* Milano, Italy: Angeli.

Sassen, S. (1991). *The global city: New York, London, Tokyo.* Princeton, NJ: Princeton University Press.

Sassen, S. (1994). *Cities in a world economy.* Thousand Oaks, CA: Pine Forge.

Sassen, S. (1995). *Losing control: Sovereignty in an age of globalization.* New York: Columbia University Press.

Saunders, P. (1986). *Social theory and the urban question.* London: Hutchinson.

Sorkin, M. (Ed.). (1994). *Variations on a theme park.* New York: Noonday.

Sudjic, D. (1993). *The 100 mile city.* London: Flamingo.

Tisdale, H. E. (1942). The process of urbanization. *Social Forces, 20,* 311-316.

van den Berg, L., et al. (1990). *Marketing metropolitan regions.* Rotterdam, The Netherlands: Erasmus University.

Vegetti, M. (Ed.). (1976). *Polis ed economia nella Grecia antica.* Bologna, Italy: Zanichelli.

Webber, M. M. (1964). The urban place and the non-place urban realm. In M. Webber (Ed.), *Explorations in urban structures* (pp. 79-153). Philadelphia: University of Pennsylvania Press.

Wells, H. G. (1962). *Anticipations: The reaction of mechanical and scientific progress on human life and thought.* New York: Harper & Row. (Originally published 1905)

Whose Britain is it? (1989, November 20). *Newsweek,* p. 54 ff.

Wolfe, T. (1968). *The pump house gang.* New York: Noonday.

9

Which New Urbanism? New York City and the Revanchist 1990s

NEIL SMITH

Back in France during the 1890s, the revanchist movement had hit its peak. *Revanche* is the French word for *revenge,* and the revanchists represented all the wholesome right-wing revenge and reaction that today would be associated with family values and militia politics, Christian fundamentalism and death penalty pugilism, patriotic bigotry and vicious racism. A movement of reactionary populists led by Paul Déroulède and organized as the Ligue des Patriotes, they bristled at the perceived liberalism of the Second Empire and the socialism of the Paris Commune and denounced the ignominy of defeat by Bismarck and the decadence and apparent impotence of the late-century monarchy. On all sides, they sensed that the sanctity of bourgeois nationalist order was threatened; a bitter revenge was to be exacted against those at home and abroad who had stolen the nation (Rutkoff, 1981). The revanchists mixed militarism and moralism with claims about public order on the streets as they flailed around for enemies.

It is not France during the 1890s that concerns me here so much as the United States a century later, and especially New York City. On August 9, 1997, after police were called to break up a fight between two women at a nightclub in the New York borough of Brooklyn, Abner Louima, a Haitian immigrant who also had tried to intercede in the fight, was arrested and thrown in a squad car. Apparently assaulted en route, Louima was hauled into the bathroom when they reached the police

AUTHOR'S NOTE: Parts of this chapter appeared in an earlier piece, "Giuliani Time," *Social Text,* Volume 57, 1998. I am grateful to Cindi Katz, Dorothy Hodgson, Ida Susser, and James DeFilippis for very useful comments and suggestions.

station, and there four white cops allegedly sodomized him with the handle of a toilet plunger and then thrust it into his mouth. Left in a crumpled heap, only later was he taken to hospital, where he was diagnosed with severe internal organ damage. Over the next 2 weeks, through several operations, Louima fought for his life. As the story emerged from enraged hospital nurses, it was widely reported that while they assaulted Louima, the police officers invoked the names of the current mayor and his predecessor, respectively, chanting "It's Giuliani time, not Dinkins time."

In this chapter, I argue that a new urban regime began emerging in the United States during the 1990s, marking a significant shift from the Keynesian urbanity installed following Franklin Roosevelt's New Deal of the 1930s. The contours of that "new urbanism" are particularly evident in New York City, still widely but erroneously conceived as a bastion of liberal power and excess, and much of the chapter relates the dramatic shift in public political presumption occurring in that city during the final decade of the 20th century. Part of the shift has to do with the altered economic position of cities in a "globalized" economy, but part of it also is more immediately political. In the vacuum created with the disintegration of liberal urban policy since the 1980s, a growing revanchism now steers social and economic change and the city's administration. This new urban revanchism expresses a decisive shift in the political geography of *modern* urban capitalism.

The scale of the shift to a new urban revanchism is not yet clear, nor is it clear how trenchant a shift it will be. It is in no way unique to New York, as I briefly illustrate, nor is it necessarily restricted to the urban scale. It should not be confused with the architectural design movement that calls itself the "new urbanism." Still, the impulse that gave rise to the nostalgic traditionalism of so-called new urban architecture, every bit a struggle to reclaim a lost bourgeois order, is shared with the new urban revanchism. After exploring some of these connections between revanchist and architectural designs for a new urbanism, I conclude with a brief consideration of alternative urban futures.

■ Giuliani Time

As the police officers faced criminal indictment, a recovering Louima later was reported to have retracted the claim that they chanted "It's Giuliani time" while assaulting him. The image, however, already was seared in the popular imagination. "Giuliani time," in some formal sense, began in early 1994 when Rudolph W. Giuliani took over as mayor of New York City, replacing David Dinkins. Dinkins had angered the city's police force by being openly critical of police behavior, and Giuliani, a former federal prosecutor, was the beneficiary as the police flocked to his support in the 1993 mayoral election. But "Giuliani time" began for real a few months later with the issuance of the innocuously named "Police Strategy No. 5" dedicated to "Reclaiming the Public Spaces of New York." Bearing the names of both the mayor and the police commissioner, William J. Bratton, this document, more than any other, marked the advent of a fin de siècle American revanchism in the urban landscape, a founding document of a different type of new urbanism (Giuliani and Bratton, 1994).

The economic depression that belatedly followed the 1987 stock market crash lasted in New York until at least 1994. The frenzied investment in the urban landscape that marked the 1980s ended abruptly, and the careless optimism hitched to the Dow Jones and the information revolution, gentrification, and consumerist abandon went the same way. With the national economy already well into a restructuring that would be captured in the language of globalization, social and economic identities were fundamentally destabilized as local economies were turned inside out. The end of the 1980s seemed to lead back to the 1970s when fiscal bankruptcy inaugurated a major retrenchment and restructuring of city services while simultaneously enhancing the role of the private sector in the city's economy. At the beginning of the final decade of the 20th century, bridges crumbled and streets went unrepaired in record numbers; landlords again began to abandon buildings; and city and state services in housing, education, and health care deteriorated or were cut back. Although hundreds of millions of dollars in tax abatement "geo-bribes" flowed regularly to attract or keep mega-corporations in the city, the official unemployment rate soared to more than 10%. Public fear replaced self-centered optimism during the early 1990s. It was this fear that incoming Mayor Giuliani skillfully appropriated.

"The downward spiral of urban decay" was real enough, but what Giuliani achieved in Police Strategy No. 5 was two things: first, a visceral identification of the culprits, the enemies who had stolen from the white middle class a city that members of the latter assumed to be their birthright, and, second, a solution that reaffirmed that right. Rather than indict capitalists for capital flight, landlords for abandoning buildings, or public leaders for a narrow retrenchment to class and race self-interest, Giuliani led the clamor for a different type of revenge. He identified homeless people, panhandlers, prostitutes, squeegee cleaners, squatters, graffiti artists, "reckless bicyclists," and unruly youths as the major enemies of public order and decency, the culprits of urban decline generating widespread fear. "Disorder in the public spaces of the city" presented "visible signs of a city out of control, a city that cannot protect its space or its children":

> New Yorkers have for years felt that the quality of life in their city has been in decline, that their city is moving away from, rather than toward, the reality of a decent society. The overall growth in violent crime during the past several decades has enlarged this perception. But so has an increase in the signs of disorder in the public spaces of the city. (Giuliani & Bratton, 1994, p. 5)

Criminality is here spatialized, even postmodernized, insofar as the signs and the symptoms substituted for the reality and certain social presences in the landscape—the litany of culprits—are made the causes of decay. Urban decline, street crime, and "signs of disorder"—the sign melded with the deeper swell of historical change—are here galvanized into a single malady. Deep-seated fears and insecurities are enlisted to fuse and conflate physical and psychic safety; the symptoms *are* the cause. Sanitizing the landscape will reverse the city's decline, opening up the possibility of a glorious new urban destiny. Revenge against the sources of disorder was raised to a moral obligation with the advent of "Giuliani time." Thus, an anti-homeless poster campaign on the subways blared over simultaneously demeaning and threatening images of homeless people—"Don't give them *your* money!" With the malefactors identified, the solution to "spiraling urban decay" was readily displaced toward an eradication of the *signs* of disorder littering the urban landscape.

This anti-crime initiative begun in 1994 was a central plank of the emerging revanchist city. The solution followed inexorably from the vengeful "logic" of the analysis. The police were ordered to pursue with zero tolerance all supposed petty criminals whose actions "threatened the quality of life." Once arrested, these individuals' cases were to be prosecuted vigorously. In addition, a "database" was established for tracking homeless people, and precinct commanders were given widely expanded powers to bypass legal and bureaucratic checks on police behavior—"proactive" policing, as Police Strategy No. 5 described it. How police cleaned up the streets was their business, Giuliani advised the New York Police Department, qualifying the carte blanche only with the caution that, of course, police ought to stay within the law.

The transfer of trust and power from political electees to the police, the dissolution of existing constraints on police power in the name of "reversing the decline in public order," and the sharp class and race delineation of the assumptions undergirding a "decent society," "urban order," and "quality of life" have spearheaded an ominous remake of the New York City streetscape. These actions also present a chilling sign of potential urban futures. Equally portentous is the unembarrassed avowal, embedded in Police Strategy No. 5, that the fortunes of New York City are centrally, even primarily, a question of police strategy. This conclusion is in no way an overinterpretation. At a 1994 forum on urban crime, Giuliani publicly bemoaned the fact that freedom had so many naive adherents, whereas "authority" got a bad rap: "What we don't see is that freedom is not a concept in which people can do anything they want. . . . Freedom is about authority," the mayor insisted. "Freedom is about the willingness of every single human being to cede to lawful authority a great deal of discretion about what you do" ("Some Free Speech," 1994). Such was the happy confusion of freedom and authority enjoyed by Louima on "Giuliani time."

Homeless people, numbering about 100,000 in New York City, bore the brunt of 1990s revanchism. The anti-homeless antagonism that smoldered during the 1980s burst into flames as official urban policy during the early 1990s. There are many other scapegoats, of course—people of color and immigrants, women and welfare recipients, working class people and gays/lesbians, and especially squatters. With the onset of economic depression at the end of the 1980s, however, public sympathy for "the homeless"—a brutal discursive reification—evaporated entirely. Homeless people, those evictees from the private and public housing markets who inhabit the barest interstices of public space, were quickly identified as the No. 1 public blight.

In its origins, if not its systematic application, the criminalization of homelessness, crystallized in "Giuliani time," actually was initiated by democratic mayors during the 1980s, especially around the struggle for Tompkins Square Park in New York's Lower East Side. Long a symbolic space of political opposition in the city and by the late 1980s a haven for people evicted from the gentrifying housing market, the park was the scene of a violent police riot in August 1988 (Smith, 1992, 1996).

On the defensive for 3 years, it was not until 1991 that the city administration embarked on a concerted geographical campaign to remove evictees from central gathering points, homeless settlements, and gentrifying neighborhoods—from the bridges, tunnels, parks, and empty lots that they occupied—and eventually to push them out of Manhattan as a whole. Whereas many Manhattan neighborhoods hosted homeless settlements in 1991, by a frigid February morning 6 years later, the Giuliani administration could boast that it had demolished the "last shantytown" in Manhattan in the railway tunnels of the old Penn Central Railroad (Kershaw, 1997). The land was being demolished for a new condominium development, "Riverside South" (sponsored by infamous real estate developer Donald Trump), and the city administration was aggressively indifferent to the fate of evictees. Actually hoping to maximize the migratory thrust of these attacks on homeless settlements, the Giuliani administration persistently refused to provide the Coalition for the Homeless with any information about the timing or locations of "homeless sweeps," thus deliberately preventing the provision of vital services.

The political geography of eviction produced more gruesome signs as well. In every previous winter since the mid-1980s, New Yorkers had become inured to the news that homeless people periodically froze to death in Manhattan parks or on benches or else died in fires as they tried to avert that fate. Now driven from the interstices of the central city, ever more desperate and no longer commanding media headlines, homeless people retreated to the outskirts where coastal scrublands, boardwalks, highway on-ramps, and the fenced-in desolation around airports (especially JFK International) offered the last secluded shelter on the urban margins. But geography is no cure for homelessness, and the need for winter heat obeys no spatial bounds in the New York outdoors. In February 1997, at least four homeless people burned to death in homeless shanties, this time under highways and boardwalks in the outer boroughs (Jones, 1997).

Police Strategy No. 5 registers the centrality of homelessness via omission. Constrained by the fact that homelessness per se is not illegal and that sufficient sympathy might persist to provoke anti-Giuliani sound bites, the mayor and his police commissioner dangled a redefinition of homeless people in front of precinct commanders and street officers—"dangerous mentally ill street people." The euphemism is revealing. It never is defined, and the range of interpretation is as broad as it is vague. Meanwhile, the city and the mayor are periodically held in contempt by the state supreme court for their refusal to follow legal requirements for sheltering evictees.

City Hall has made no secret of the fact that this and other policies are explicitly designed to rid New York City of homeless people, regardless of where they end up. Planned shrinkage, the 1970s doctrine devoted to reducing public expenditure, is back in monstrous form, with the targets now clearly identified as neither budgets nor infrastructure but rather people themselves. Shrinkage of the poor population in general, including homeless people, is "not an unspoken part of our strategy," the mayor once explained at a "confidential" meeting of newspaper editors. "That is our strategy" (quoted in Barrett, 1995).

The strategy has worked as erstwhile liberals have applauded. "We're not suicidal liberals anymore," announced one community activist turned anti-homeless crusader (quoted in Goldberg, 1994). Or, in the true spirit of revanchism,

> As a lifelong New Yorker doing graduate studies in Chicago, I am grateful that New York reelected Rudy Giuliani.
>
> Of course he's a dictator, but maybe it takes a dictator to run this city. The streets are safer, there's hardly any porn on 42nd Street (believe me, I've looked!). And people actually want to live in New York City.
>
> Sure, he's unfair to the poor, but what politician isn't? At least Rudy isn't phony about it. ("Rudy Rules," 1997)

The unsurprising result of the city's anti-homeless and anti-poor policies has been to place more people on the streets, whether forcibly (as with the parallel campaign to evict squatters) or less directly. Indeed, a neat confluence of vengeful motives has led the Giuliani administration to terminate city contracts with several neighborhood organizations providing housing for hundreds of homeless people. Among these are the Harlem Restoration Project, an activist neighborhood organization that has been very successful in renovating Harlem tenements for poor residents, and Housing Works, an activist organization devoted to housing 260 people with AIDS who otherwise would face homelessness. In both cases, these provider organizations also had vociferously challenged Giuliani's policies, and an act of petty mayoral revenge against political opponents dovetailed with revenge against homeless people themselves.

Parallel with public sector attacks on housing provision for the city's most marginal residents, private sector landlords and developers in

1997 launched an unprecedented campaign against city rent control laws. Approximately one third of New York City's 3 million residential units are covered by some form of rent regulation that limits the level of annual rent increases to preset levels. With large contributions to pivotal state and city politicians and a large budget for advertising and campaigning, landlord and developer associations began lobbying in 1997 to eliminate rent protection for poor tenants entirely. Although last-minute tenant and community activism succeeded in blunting this assault and largely sustained legislation aimed at protecting the poorest of the working class from economic or legal eviction, the landlords' intent dovetailed with public policies aimed at dispersing poor people from the city.

Whereas the different threads of the new urban revanchism came together by 1994, that also was the year in which New York began to emerge from the depths of recession. Wall Street led the charge, but real estate was quick to follow. Languid markets in many Manhattan neighborhoods suddenly sprang back to life—SoHo, newly fashionable Chelsea, the Upper West Side, even Inwood. Apartments that could not be given away during the early 1990s now routinely attracted bidding wars. The unprecedented feeding frenzy in Tribeca real estate, adjacent to the downtown financial district, was likened in local newspapers to "truffle pigs" following "the law of the market" (Sifton, 1997). The Lower East Side on the other flank of the financial district, the crucible of 1980s avant-garde art and a symbol of hip gentrification as well as the location of most intense anti-gentrification struggles, embraced the surge of reinvestment and gentrification with a Web site soap opera, a record label, and a clothing line all exploiting the neighborhood's name. "We've tried to brand-name the Lower East Side," said one enthusiastic entrepreneur (quoted in Gabriel, 1996). Churches in Harlem, now a top international tourist attraction, have had to regulate admission by mostly European and Japanese tourist voyeurs (white Americans presumably remain too squeamish) and to initiate detailed rules of behavior to allow unfettered worship—"no flashes during prayers."

Amid the gentrification frenzy of the 1980s, Paul Berman once queried why it is "the strange tenacity of its ancient neighborhoods" that "governs New York creativity" (Berman, 1988). The determinism, he concluded, "is geographical. Ghosts run the city. Or why is it in the New York arts, if a building isn't seedy, it can't be taken seriously? Why else do the arts prosper only in certain downtown streets, never in the fancy neighborhoods?" A neighborhood with its own Web site soap opera might be suburban cool, but it no longer is edgy. So, in addition to filling in the previously capitalized neighborhoods of Manhattan and inner Brooklyn, this new wave of gentrification "is marching to the fringes." Brooklyn Heights and Park Slope already were targeted during the 1980s, and Cobble Hill and Carroll Gardens were targeted less intensely, but all were experiencing the resurgence of gentrification by the late 1990s. Other neighborhoods, such as Greenpoint and the Old Brooklyn area between the Manhattan and Brooklyn bridges, are witnessing sustained gentrification for the first time. The locus of artistic culture also migrated outward during the early 1990s to Williamsburg and Long Island City, and by the end of the decade, these neighborhoods, in turn, were facing gentrification. Yet, the quintessential sign of New York gentrification during "Giuliani time" came at the city's center. While avant-garde art prepares the ground for reinvestment at the edge of the inner city, a political and economic campaign to remake the core spearheads the Disney-sponsored suburbanization of Times Square.

The contradictions of economic recovery go much deeper than the sprouting of odd symbolic juxtapositions in the new urban landscape. Rents soared after 1996, and more and more people again found themselves living on the edge. As the real estate markets soared, landlord brutality aimed at evicting tenants also returned with a literal vengeance. In one case reminiscent of the late 1980s, a Brooklyn landlord was charged with having organized a "campaign of terror" to clear a building for highly profitable redevelopment. Attempted murder and arson were the charges; the landlord was accused not only of burning his own building to clear the tenants but also of injecting one tenant with a drug overdose (Sullivan, 1997).

The return of economic delirium to the city, ever more slickly packaged, comes without one central ingredient of the previous boom: Whereas the number of homeless people in the city has increased an estimated 15% between 1994 and 1997, there has been no resurgence whatsoever in public sympathy for homeless people, and drastic cuts have decimated the facilities and services that assisted homeless people or insulated others from becoming homeless. The gathering revanchism of the early 1990s might have been provoked by economic recession, but mid-1990s prosperity did nothing to staunch the cries for revenge. Curious as it seems, the avarice and venality of the 1980s does actually appear to have been a kinder and gentler urbanism than the chilling 1990s.

I have focused on New York here, but the new urban revanchism is neither restricted to Gotham nor even necessarily at its most vengeful there. By 1997, more than 70 cities in the United States had enacted legislation banning or severely circumscribing panhandling, sleeping in public places, and other behaviors associated with homeless people. The most sweeping of these laws have been challenged, often successfully, on grounds of free speech or excessive punishment, but some municipalities have been imaginative in circumventing challenges. Unable to ban panhandling outright, New Orleans sought instead to license it, requiring panhandlers to pay $50 for the privilege of begging. The intent of these laws is the selective privatization of public space, whereas spaces that remain truly public are subject to increasingly intense surveillance (Sorkin, 1992). Nor is this shift restricted to traditionally conservative cities. San Francisco, Berkeley, and Santa Monica all have significantly reversed previously liberal policies concerning the public presence of homeless people. Here, for example, is a description of the "vagrant policy" in Santa Ana, a Southern California city, offered by a city bureaucrat:

> [The] city council has developed a policy that the vagrants are no longer welcome in the City of Santa Ana. . . . In essence, the mission of this program will be to move all vagrants and their paraphernalia out . . . by continually removing them from the places that they are frequenting in the city. (*Tobe v. City of Santa Ana,* 1994, pp. 386-388)

These and other legal initiatives comprise what Don Mitchell calls a strategy of "the annihilation of space by law" (Mitchell, 1997).

The Revanchist 1990s

The new urban revanchism is, in many ways, tied to the shifting niche of cities in the global economy. There is a lot of truth to the contention that whatever other myriad functions and activities it housed, the late-19th- and 20th-century capitalist city is geographically defined as the locus of social reproduction. Keynesian urban policy from the 1930s to the 1970s was devoted to the broad-based subsidy of local social reproduction that underscored capital accumulation in economic, political, and ideological terms. From Henri Lefebvre, to Manuel Castells, to David Harvey, urban theorists understood the so-called urban crisis of the 1970s as emanating from a crisis of social reproduction provoked by the dysfunctionality of racism and oppression and the contradictions between an urban form constructed according to strict criteria of profitability yet simultaneously required to facilitate the reproduction of a labor force (Castells, 1977; Harvey, 1973; Lefebvre, 1976). Nearly a quarter century later, amid the white heat of globalization, these diagnoses seem almost quaint; the urban scale has been significantly unhinged from such definitive responsibility for social reproduction.

There are several dimensions to this shift. First, the erosion of national boundaries as economic barriers to capital mobility has dramatically expanded the range of capitals that are free to move to where lower costs of social reproduction, in turn, lower the costs of production. Second, the unprecedented migration of labor since the mid-1970s has increasingly distanced local economies from automatic dependency on local labor. There might not yet be a global labor market, but there is global interdependence of labor markets (Castells, 1996, p. 239). Third, forced into a more competitive mode vis-à-vis capital and labor, local states (including city governments) have offered carrots to capital and applied the stick to labor. They have become much more selective about the extent and level of subsidy for local social reproduction because they can depend, to a greater extent, on imported labor whose costs of reproduction are borne elsewhere. Fourth, the same pressures applied to the national state have led to a dramatic erosion of social capital provision on that scale as well, intensifying the pressure on city governments to further sever their responsibilities for social reproduction. Fifth, amid the restructuring of production beginning during the 1970s and with class- and race-based struggles broadly receding, city governments had an increased incentive to abandon that sector of the population surplused by both the restructuring of the economy and the gutting of social services. Comparatively low levels of class and community struggle prior to the uprising were crucial in encouraging the virtual nonresponse by government to the 1992 Los Angeles rebellion. In dramatic contrast to the ameliorative policies that sprouted after the Watts and subsequent uprisings during the 1960s, policies designed to open a safety valve on searing social tension, the response after 1992 comprised heightened repression barely disguised by minor reconstructive palliatives.

No automatic response to economic ups and downs, this visceral revanchism is fostered by the same economic uncertainties, shifts, and insecurities that permitted the more structured and surgical abdication of the state from the throne of social reproduction of labor. Revanchism is in every respect the ugly cultural politics of neoliberal globalization. On different scales, it represents a response spearheaded from the standpoint of white and middle class interests against those people who, they believe, stole their world (and their power).[1] This was the theme of Tom Wolfe's brilliantly twisted *Bonfire of the Vanities* back in 1987, and it has been picked up elsewhere (Wolfe, 1987). More recently, *Harper's* editor Lewis Lapham connects a turn toward "reactionary chic" with the bitterness of downward mobility experienced by many in the upper middle class (Lapham, 1995). Unemployment, generally associated with the industrial working class, has transmogrified during the 1990s, along with shifts in the structure of the U.S. economy, into an increasingly white-collar experience. In the words of one liberal turned bitter revanchist, "The two halves of Sixties liberalism—social license and economic restrictions—reinforced each other. Rent, zoning, and business legislation" left the large cities "saddled with expensive and inefficient governments" benefiting only the poor who had failed to compete and social workers and "caring" professionals who "came to live off the personal failings of the big cities' dependent populations" (Siegel, 1997, p. xi).

Domestically, then, revanchism is allied with the assault against so-called political correctness. Both galvanize a revenge and reaction against imputed enemies who stole the country, a theft located in the emergence of 1960s radical politics—socialists and feminists, anti-racists and environmentalists, opponents of the war in Vietnam. Internationally, it represents an attack against enemies of the "American Century" for frustrating a long-term manifest American destiny. Defeats in Vietnam, Nicaragua, and (especially) Iran provided political and racial profiles of ready enemies, but as U.S. military commitments in the Middle East proliferated after the 1970s, defending the oil trade and Israeli expansionism, and as ideological certainty grew more precarious after the 1989 implosion of official communism in Europe and Central Asia, that profile has narrowed. Arabs have become the sacrificial targets of official U.S. revanchism on the international scale. The sacking of Baghdad during the 1990s by two different U.S. presidents bears all the hallmarks of revenge in the name of U.S. interests.

The Triumph of Meanness (Mills, 1997), to cite the title of Nicolaus Mills's book, has occurred on all of these scales, from the local to the international, yet it is hardly an example of "America's war against its better self" (the book's subtitle), which assumes the same nationalist ground as the meanness it challenges. Even less is the new revanchism a casual deviation from liberal American normalcy, requiring a reinfusion of liberal decency. American liberalism of the 20th century was the exception, itself a deviation and not the rule. The new revanchism represents a recrudescence of precisely that fundamentalist 18th-century liberalism on which the United States was founded, which, in a progressive move for the time, championed the connection between democracy and individual self-interest. Smith, Hobbes, Locke, Hume, and Kant were its progenitors, and neoliberalism, "reactionary chic," the new conservatism, the revanchist city, and even the current Kant revival in social theory are its direct descendants. As critics around the world understand, the decline and fall of Keynesian interventionism—from the welfare state and urban policy to state-centered models of international development—marks a structured political adjustment toward what is widely understood as neoliberalism.

Two alternatives have invaded the vacuum created by the strategic withdrawal of the national and local state from the management of social reproduction. On the one hand, a market more powerful and extensive than ever before is increasingly endowed with the power to establish social norms; the market is increasingly the determinant of "natural" social relations and consequences. These market assumptions suffusing social relations are pervasive. They are expressed in "truffle pigs" in the real estate market but also in the economists' self-flattery that a "natural" unemployment rate exists. They steer interpretations of life emerging from the Human Genome project and the new Malthusianism, and they have strangled most radical life from an environmentalism gone corporate. But the marketization of everything is matched by a second response insofar as the withdrawal of the state is selective as well as strategic: The crisis for those marginalized by the new capitalism also is increasingly policed.[2] Already turned upside down by the sweeping economic restructuring that began during the 1970s, the disruption of social reproduction unfolds more fully at the century's end (Katz, 1997).

The policing of the crisis has only just begun. Precise estimates of the social effects of President Bill Clinton's 1996 welfare "reforms" are difficult to make, not least because the different provisions are to take effect at different times over the next 5 years and because the response by local state and city governments is not clear. But the discontinuance of numerous federal programs, the displacement of responsibility to local and state governments, and the coupling of welfare benefits to city work programs—"workfare"—already have significantly reduced welfare rolls while providing no significant alternative (Finder, 1998; Greenhouse, 1998; Swarns, 1997; Toy, 1998).[3] The Giuliani and Santa Ana cases already suggest the likelihood of a geographical bidding war—the "race to the bottom"—where local authorities bid down their service provision in a vain attempt to repel homeless people, the unemployed, sick and disabled people, young people in need of education, and others dependent on social services toward less austere states and municipalities. According to one estimate, for Los Angeles alone, the 1996 welfare reforms will have the following effects over the next 5 years: Some 16,000 to 227,600 people will lose medical insurance, 7,400 to 30,000 women will lose prenatal care, 8,800 to 15,400 disabled people will lose domestic support services, and child abuse will rise demonstrably. In addition, as many as 50,000 jobs could be lost as a result of direct and indirect economic effects, and the number of people experiencing homelessness could rise by 190,000 (Wolch & Sommer, 1997).

Even if such estimates prove high and many victims of welfare reform succeed in scratching out livings at or above subsistence levels, a large additional surplus population nonetheless is in the making. Whereas during the late 1960s the response to the crisis of social reproduction involved a massive infusion of funds aimed at buttressing standards of living for the marginally employed and at appeasing opposition, today all the signs point to repression rather than buy-off as the strategy of control. The discrepant responses to the Watts uprising of 1965 and the South Central uprising in 1992—the first attracting a flurry of new urban programs, the second attracting a full-scale arming of a police force whose brutality against Rodney King was, in any case, the source of the uprising—symbolize the change. The massive expansion of the prison population and of prison construction, from California to Texas, tells us that this solution already is in place (Gilmore, 1997).

Of course, fear of crime is not just a middle-class white issue, even if it is this latter spin on crime that organizes the political response. The decimation of more traditional means of social reproduction and work has spawned a range of alternative economies, from street mugging and burglary to prostitution and the drug industry, which disproportionately affect working class and poor neighborhoods. In Harlem or South Central, however, there is no automatic assumption that the police are the appropriate response to high crime rates or even that police presence increases public safety. As the King and Louima cases highlight, police brutality has become epidemic during the 1990s in neighborhoods where as many as half of the male youths are present or past "clients" of the prison system. Yet, it is the threat felt by the well-off white middle class rather than neighborhood appeals for jobs, housing, and education that has scripted the official response to crime rates that are, in any case, declining in many cities.

Another neat confluence of interests occurs here. The most extensive prison-building spree in the world, beginning in California in 1982, was fueled not by dramatic increases in crime insofar as crime rates actually were decreasing throughout the period. Rather, surpluses of capital, land, and population (white, black, and Chicano youths whose educations left them unprepared for the new economy), together with the burgeoning illegitimacy of the state in the eyes of voters, posed an intense crisis during the early 1980s. The resulting crisis found a convenient solution in a massive prison-building program. Surplus land, capital, and people all were consumed while the state won much-needed approval in the burgeoning suburbs of populous Southern California (Gilmore, 1997). Revanchism works. And the revanchist city is the place where both alternatives to state abrogation, the market and intensified policing, operate simultaneously and most intensely, welded into a proximate postmodern politics of sign production and eradication.

Invoking the fear that fascism was just round the corner became the object of humorous scorn on the left during the 1970s, much as political correctness began life as a source of comedic ridicule also internal to the left. With less of a left about and with the fulcrum of public discourse veering rightward, it has fallen to neoconservatives to raise the same concerns. None other than Edward Luttwak, conservative prophet of global economic competition, is now warning that fascism could be "the wave of the future" (Luttwak, 1995). Whether true or not, the fact that the right has taken over the left's rhetoric, and means it sternly, is cause enough for concern about the political future.

■ Which New Urbanism? Geography as Revenge

As the new urban revanchism began to emerge during the late 1980s, there arose quite separately a movement in urban architectural design known as the "new urbanism." Steeped in a critique of monolithic modernist design and suburban strip mall development oriented to automobile access, the new urbanism proposed to construct a new generation of suburbs replicating the dense scale of the walking city and the nostalgic architecture of various bygone eras and places. Initially led by architects Andres Duany and Elizabeth Plater-Zyberk, whose Seaside resort in the Florida panhandle is widely seen as the prototype new urbanist project, the movement matured in 1993 into an organization adopting the expansive title "Congress for the New Urbanism." New urbanist developments now dot the country, and coalitions have been established with, on the one hand, Prince Charles's crusade for a more human, heritage-conscious architecture in Britain and, on the other, Disney's community engineering project at Celebration, Florida. If Herbert Muschamp is to be believed, then the new urbanism is "the most important phenomenon to emerge in American architecture in the post-cold war era" (Muschamp, 1996, p. 27).

Architectural new urbanism and the new urban revanchism would seem to be quite disconnected developments dealing with quite different worlds. That is precisely the point of new urbanist design. The picture postcard Victoriana of Seaside, replete with white picket fences, pedestrian-friendly streets and alleys, chartreuse features, seaside gazebo style, and unrelieved middle class congeniality, not only exudes a one-dimensional vision of the urban future dressed up as nostalgic recall but also rigidly polices the present. Developed along a silver beach in a county of the Florida panhandle that has a large minority of black residents, Seaside is stylistically as much as statistically white and wealthy. Very few nonwhite faces venture into town, although Andres Duany, one of its primary designers, is eager to point out that some of his best architects were black.[4] An agency manages "cottage rentals," each with its own name, for as much as $4,500 per week (six bedrooms), whereas at the low end, one-bedroom cottages can be rented for $277 per night ($242 in the off-season).

There is no mystery about for whom this new urbanism is built. The design styles distill the most traditional social assumptions of gender, class, and race. Dripping like candle wax with sentimentality for "the human scale," a mushy metaphor that hides more than it reveals, Seaside openly and exuberantly celebrates the 19th-century urban ideal of yeoman New England. The past evoked in the promise of a new urban future is the narrowest and most elitist of founding fantasies, and the resulting landscape naturalizes a wide plank of privileged presumptions of the social norm. In its discreetly bounded single-family homes, assumptions about gender roles are as neatly kept up as the postage stamp–sized gardens. As you enter Seaside, there is no sign on the road to say "No Irish need apply" or its less verbalized 20th-century equivalents—"If you're black, stay back," "If you're working class, be out by five," "Women in the kitchen, please." The design style already speaks the exclusionary message with delicate, handkerchiefed smugness.

Celebration exudes much the same faux-populist elitism. Designed and rigidly controlled by Disney, Celebration is theme park living without the rides and amusements. The houses come in six styles: Colonial Revival, Mediterranean, French ("returning veterans from World War I were intrigued by French country houses"), Victorian, Coastal, and Classical (with its pillared grandeur, it would be more honestly described as "Plantation"). Unlike other Disney theme parks, however, the Disney presence is not ostentatiously advertised. The town center features numerous shops around a pond and water fountain, but the main square is dominated by a four-story tower overlooking the whole development, a monument to real estate voyeurism, so much so that at the base of the tower, where a tourist information office might reasonably be expected, sits Celebration Realty, clearly the fulcrum of the town's weekend activity. It easily outshines the adjacent town hall, itself privatized. Exhibits and a video presentation in Celebration Realty steer prospective customers toward homes on offer. In the spring of 1998, a few studio apartments in the center sold for as low as $167,000, but the theme houses averaged between $450,000 and $600,000, with some over $1 million. In Celebration Realty, there is a rare hint of the hidden hand of Disney; on the door where one typically might read "employees only," a sign says "cast members only." Such a merger of urban design, architecture, and the state apparatus into a "single comprehensive security effort" represents what Mike Davis describes, in a different context, as "the bad edge of postmodernity" (Davis, 1991, p. 224). But in this case, it represents the bad edge of super-concentrated modernity.

The new urbanism is different things in different places. Nonetheless, the evolving crossover between the style and politics of the new urbanism and the new Disney landscapes of the urban ideal should cause all of our critical ears to prick up in search of any and all political signs. So long as Disney landscapes accumulated the obviously exotic past and distant present as a means of transporting the "customers" through time and space (Fjellman, 1992), the distance between fact and fantasy always was palpable. But the collapse of time-space between Disney recreations and the world it purports to romanticize, most viscerally in "Old Key West" at Sea World just 375 miles from the real thing, shovels all of the investment of Disney fantasy into the new urbanism. The traffic of social and architectural signs between Disney and urban planning reaffirms a future past that is violently exclusionary in the present. It is fitting, therefore, that the makers of *The Truman Show,* a darkly troubling film portraying a life of total, happy surveillance, chose Seaside's new urbanism as the appropriate landscape of a dubious future.

What is the connection between this new urbanism and the revanchist city? At its most obvious, the new urbanism represents an escape from the city, but its significance goes deeper. The glaring white landscapes of Seaside are umbilically tied to the revanchist city as scripts of an alternative source of power—of power regained. New urban landscapes are fantasies that the world might be different, fantasies that, within the rigidly controlled spaces of new urban developments, are made real. Here, within the design concepts and site plans of the new urbanism, the world can be made safe for a self-conscious liberalism that elsewhere recoiled into revanchism. Screaming that Seaside is *not* the revanchist city, the anxiously relaxed, self-referential traditionalism of its white picket fences conceals its mirror image. The revanchist city is the alter ego of the new urbanism, its Frankenstein (the monster of its own making), the Marx to its very own Derrida (Derrida, 1994). The revenge of the new urbanism is not the visceral revenge of New York streets and politics. It is a revenge hardwired into the institutional control of the landscape and its spatial location. Precisely in its escapism, the new urbanism posits geography as the means of revenge.

Without the revanchist city, the new urbanism has no rationale; the past it evokes has no future except perhaps for a small elite. It is not a solution to, but rather an accomplice with, the revanchist city, expressing revenge in the apparent silence of space and delicacy of urban design rather than the declamatory rhetoric of Giuliani. But the revanchist city, rather than Seaside's picket fences, shapes most people's experience of a new urbanism. If we are to take seriously the inevitable task of prescribing and enacting a new urbanism different from the models that currently prevail, then it is important to begin to consider alternatives to a deepening revanchism.

■ Alternative New Urbanisms

Scandalizing the revanchist city is important if we are to get a clear sense of how the political wind is etching and eroding the urban landscape. It also is important to retain an international perspective. If revenge against the poor seems a novel development in U.S. urban policy, then it is vital to remember that in many other places, from the Rio de Janeiro *favelas* (where police squads murdered street kids) to the vengeance of ethnic cleansing in Sarajevo, it is a familiar feature of daily life and can be much more brutal. It also would be a mistake to confuse repression with revanchism, however much the latter might include the former. Repression may have many rationales, whereas revanchism is about revenge.

Identifying and scandalizing revanchism is, however, just a first step. Revanchism is not a politically or socially necessary outcome of economic crisis and restructuring, nor is it the automatic political or cultural progeny of the morning after 1960s "liberalism." It was a distinct choice by a narrow group of political and corporate leaders who really do constitute, with others, a ruling class. In New York City, the foundations were laid with the corporate takeover of the city's administration after the mid-1970s fiscal crisis, but the repressive cutbacks of that period did not evolve into a clear revanchism until the end of the 1980s. That revanchism is a choice is illustrated by one simple fact. Since 1996, the city has enjoyed an annual budget surplus as high as $3 billion, yet the mayor chose to amplify his demonization of homeless people rather than use the available funds to provide housing. This was well understood in the myriad protests that have met each of the mayor's forays against civic rights; many poor communities have demonstrated against police brutality, taxi drivers struck for a day, squatters have physically defended themselves against eviction, and so on. Although these have not defeated the strategic malice of the revanchist city, they have circumscribed it. They express the recognition that the vengeful amalgam of market and police power does not exhaust the menu of social choices and that revealing the imminent scandal of revanchism is a means of priming rather than occluding alternative urban futures.

In the search for other urban futures, we can start with the economy. Planetary neoliberalism has revealed more starkly than ever the contours of global capitalism even as the apparently stifling power of globalization and the erosion of popular political economic knowledge render the signs less comprehensible. The neoliberal remake moves capitalism strangely closer to, rather than more distant from, the forms of exploitation and oppression that Smith and Ricardo recognized, Marx subjected to biting critique, and traditional economists displace as beyond their disciplinary concern. When one has agreement on this point from *The New Yorker,* which in 1997 declared that reading Marx will give Wall Street brokers a clearer picture of how capitalism works than will all the editions of Samuelson's *Economics,* it surely is time to take Marx seriously again (Cassidy, 1997).[5] A good place to start might be Marx's analyses of economic crisis (Marx, 1867/1975, chap. 25), which offers more rigorous insights into the global economic disaster unfolding since the late summer of 1997 than do all the economists' simplistic language of "corrections" or, worse, the muddle-headed and vaguely racist euphemisms of "contagion" and "Asian flu."

If the revanchist city is our concern, then understanding the state is vital, and Marx might be less helpful here. A lot of political time has been spent since the 1970s defending different facets of the welfare state from government predation, and rightly so. In the process, however, we seem to have forgotten the critiques of the state—its class, race, and gender embodiments as well as its power—that once fueled political opposition to the capitalist state.[6] Today, virtually no public alternative counters the profoundly mistaken assumption that the left venerates the state as the appropriate agent of social change and equity. At the very least, it is time to revive and revise these critiques of the state and to reestablish a clear critical distance from the state and state policies. We might prefer to hold our noses and admit that the Keynesian state is better than its absence, but the increasing transparence of the new state as a vehicle of revenge provides a great, if lamentably idle, political resource for leveraging different ideas of a new urbanism.

The homily that "all politics is local" is fatuous and self-defeating in the face of global neoliberalism. Not that the local is irrelevant—far from it—only that it is not exclusionary; "act and think locally *and* globally" is a far better guide to action. Establishing the practical and theoretical connection between the local and the global, and all scales in between, might be the most vital political project insofar as it encourages the coalition of political visions across scale rather than fragmenting them. Central is the need to recapture the sense that a different global is possible. The interval from Jim Morrison's 1967 "We want the world and we want it now!" to the 1980s Band Aid song *We Are the World* marked a two-decade leap of idealist self-abnegation in political culture, leavened by an imperial assumption that cast our local selves as effortlessly and inherently global. Although that might be metaphysically true, it is politically a lie. In any case, that shift pales in comparison to the distance traversed by the cultural economy during the past 10 years.

Today, as the shine of identity politics from the 1980s and early 1990s begins to fade, it is important to retain a sense of the profound range of social differences that inscribes the politics of the revanchist city and, equally, the motivation for alternatives. The problem today in many places is not that there is *no politics* but that there is no concerted political *movement;* the citizens of the revanchist city are plenty political even in abstention from choiceless official elections, however much that politics largely defies organized expression. It is equally important to retain a sense of how alternatives might emerge, where the sparks of change might first fly. In the United States, the resurgence of union organizing during the late 1990s, the success in rapid succession of large strikes at United Parcel Service, General Motors, and Northwest Airlines; the reforging of links between academics and labor unions that were systematically smashed during the McCarthy period; the emergence of a new activist unionism around the Union of Needletrades Industrial and Textile Employees; and an unprecedented internationalism in union organizing provoked by the North American Free Trade Agreement and globalization all are optimistic signs of a major political realignment.[7] Equally portentous are the "living wage" campaigns that have galvanized welfare recipients and public service workers to beat back attempts by city and state governments to hire workfare recipients at the expense of unionized wage workers. Environmental justice organizations, community reconstruction movements fighting gentrification, AIDS activist groups, and many other struggles provide a glimpse of alternatives. So too did the angry 10,000-person march from Brooklyn to City Hall denouncing the mayor and police brutality in the wake of the assault on Abner Louima.

These all represent "local" U.S. crucibles from which alternatives to new urbanisms can sprout. Other locals have their own crucibles of change, from the French movement of the unemployed to the spontaneous Indonesian eruptions that dethroned President Suharto, and have their own potentially new urbanisms. Nonetheless, a happy ending to the revanchist city story will take a lot of political work and a lot of analysis coming together with a lot of vision. With liberalism "dominant but dead," to appropriate Habermas's famous diagnosis of modernity, the field is wide open and the possibilities are immense. As the Los Angeles uprising of 1992 demonstrated, top-down revanchism will not go unchallenged.

The revanchist city is at root contradictory. Although Giuliani won a second-term victory over a Democratic challenger derided in a vicious press campaign as a tired liberal who should "get over" her excessive concern for the dispossessed (cf. Apple, 1997), he was reelected despite a low voter turnout, an unenthusiastic electorate, large drops in his approval ratings prior to the election, and a widening gulf of support and opposition in terms of race and, to a lesser extent, class. More to the point, on the eve of the election, 81% of New Yorkers thought that police brutality in the city was a major problem (Nagourney, 1997). The lack of organized, citywide political alternatives is most startlingly revealed in the fact that many, if not the majority, of these people must have voted for a victorious Giuliani anyway.

Top-down revanchism inevitably invites a bottom-up response. It could take the form of the Los Angeles uprising or the more ugly form of the Oklahoma City bombing, or it could emerge from amid the widespread anger and cynicism toward all forms of government and authority that crystallizes most brittlely in the militias. Whether top-down or bottom-up, revanchism surely is not the answer. A very different type of new urbanism is the prize for organizing an alternative.

NOTES

1. I should be clear here that I do not intend some narrow identitarian explanation of political persuasion. I say the "standpoint of white and middle class interests" because the expressed interests of that group have represented the power base for revanchist politics. Many among the white middle class abhor this revanchism, and many others outside it are sympathetic. It draws in significant numbers of the white working class and some among the black middle class as well as many immigrant entrepreneurs. But the diversity of its expression should not blind us to the social origins of this revanchism.

2. For an early anticipation of "policing the crisis" in a British context, see Hall (1978).

3. Apparently without understanding the historical precedent, New York City's commissioner of the Human Resources Administration defended workfare programs on the grounds that "work makes you free." Naturally, he was stunned and embarrassed when it was pointed out to him that this was the slogan adorning the entrance arches to Nazi concentration camps—"Arbeit macht frei" ("City Official Is Sorry," 1998).

4. See Duany's marginal comments in Smith (1993).

5. The significance of this article lies largely in its venue. *The New Yorker* traditionally is seen as a highbrow magazine with a literary bent that aspires to establish *haut bourgeois* taste.

6. See, for example, Holloway and Picciotto (1978), Miliband (1973), and O'Connor (1973).

7. A major 1996 conference initiated the new coalition of labor and academic activists and was followed by the formation of "Scholars, Activists, and Writers for Social Justice." On international organizing, see Johns (1994, 1998).

REFERENCES

Apple, R. W. (1997, October 26). Heroic, or just quixotic: Messinger tilts in populist tradition. *The New York Times,* pp. 31, 34.

Barrett, W. (1995, May 9). Rudy's shrink rap. *The Village Voice,* p. 11.

Berman, P. (1988, March 15). Mysteries and majesties of New York. *The Village Voice,* pp. 32-33.

Cassidy, J. (1997, October 20-27). The return of Marx. *The New Yorker,* pp. 248-251.

Castells, M. (1977). *The urban question.* London: Edward Arnold.

Castells, M. (1996). *The rise of network society.* Oxford, UK: Blackwell.

City official is sorry for remark. (1998, June 27). *The New York Times,* p. B3.

Davis, M. (1991). *City of quartz.* London: Verso.

Derrida, J. (1994). *Specters of Marx.* New York: Routledge.

Finder, A. (1998, April 12). Evidence is scant that workfare leads to full-time jobs. *The New York Times,* pp. 1, 30 (Metro).

Fjellman, S. (1992). *Vinyl leaves: Walt Disney World and America.* Boulder, CO: Westview.

Gabriel, T. (1996, August 11). The East Village: In again. *The New York Times,* pp. 39-40 (Styles).

Gilmore, R. W. (1997). *From military Keynesianism to post-Keynesian militarism.* Ph.D. dissertation, Department of Geography, Rutgers University.

Giuliani, R. W., & Bratton, W. J. (1994). *Police Strategy No. 5: Reclaiming the public spaces of New York.* New York: Office of the Mayor.

Goldberg, J. (1994, April 25). The decline and fall of the Upper West Side: How the poverty industry is ripping apart a great New York neighborhood. *The New York Times.*

Greenhouse, S. (1998, April 13). Many participants in workfare take the place of city workers. *The New York Times,* pp. A1, B6.

Hall, S. (1978). *Policing the crisis.* London: Macmillan.

Harvey, D. (1973). *Social justice and the city.* London: Edward Arnold.

Holloway, J., & Picciotto, S. (Eds.). (1978). *State and capital.* London: Edward Arnold.

Johns, R. A. (1994). *International labor solidarity: Space and class in the U.S. labor movement.* Ph.D. dissertation, Department of Geography, Rutgers University.

Johns, R. A. (1998). Bridging the gap between space and labor: U.S. worker solidariy with Guatemala. *Economic Geography, 74,* 252-271.

Jones, C. (1997, February 12). Where three died, a home on the margins of society. *The New York Times.*

Katz, C. (1997). Disintegrating developments. In T. Skelton & G. Valentine (Eds.), *Cool places: Geographies of youth culture* (pp. 130-144). New York: Routledge.

Kershaw, S. (1997, February 28). Police remove encampment of homeless. *The New York Times.*

Lapham, L. (1995, March). Reactionary chic. *Harper's,* pp. 31-42.

Lefebvre, H. (1976). *The survival of capital.* London: Alison & Busby.

Luttwak, E. (1995, November 2). Turbo-charged capitalism and its consequences. *London Review of Books.*

Marx, K. (1975). *Capital.* Moscow: International Publishers. (Originally published in 1867)

Miliband, R. (1973). *The state in capitalist society.* London: Quartet Books.

Mills, N. (1997). *The triumph of meanness: America's war against its better self.* Boston: Houghton Mifflin.
Mitchell, D. (1997). The annihilation of space by law: The roots and implications of anti-homeless laws in the United States. *Antipode, 29,* 303-335.
Muschamp, H. (1996, June 2). Can new urbanism find room for the old? *The New York Times,* p. 27 (Arts & Leisure).
Nagourney, A. (1997, March 12). Poll finds optimism in New York, but race and class affect views. *The New York Times,* pp. A1, B4.
O'Connor, J. (1973). *The fiscal crisis of the state.* New York: St. Martin's.
Rudy rules. (1997, November 18). Letter to the editor. *The Village Voice,* p. 6.
Rutkoff, P. M. (1981). *Revanche and revision: The Ligues des Patriotes and the origins of the radical right in France, 1882-1900.* Athens: Ohio University Press.
Siegel, F. (1997). *The future once happened here.* New York: Free Press.
Sifton, S. (1997, March 5-11). Tribeca in transition: Truffle pigs and the law of the market. *New York Press,* pp. 1, 30.
Smith, N. (1992). New city, new frontier: The Lower East Side as wild west. In M. Sorkin (Ed.), *Variations on a theme park* (pp. 61-93). New York: Hill & Wang.
Smith, N. (1993). Reasserting spatial difference. *ANY, 1,* 22-23, 29-37. (Architecture New York)
Smith, N. (1996). *The new urban frontier: Gentrification and the revanchist city.* New York: Routledge.
Some free speech on mayor's words. (1994, March 17). *The New York Times,* p. B3.
Sorkin, M. (Ed.). (1992). *Variations on a theme park.* New York: Hill & Wang.
Sullivan, J. (1997, October 3). Landlord is charged with waging campaign of terror. *The New York Times,* pp. B1, B3.
Swarns, R. L. (1997, October 20). 320,000 have left welfare, but where do they go from here? *The New York Times.*
Tobe v. City of Santa Ana, 27 Cal. Rptr. 2d 386-8 (Cal. Ct. App. 1994).
Toy, V. S. (1998, April 15). Workfare rules used as way to cut welfare rolls. *The New York Times,* pp. A1, B4.
Wolch, J., & Sommer, H. (1997). *Los Angeles in an era of welfare reform: Implications for poor people and community well-being.* Report for Southern California Inter-University Consortium on Homelessness and Poverty.
Wolfe, T. (1987). *Bonfire of the vanities.* New York: Farrar Straus Giroux.

10

Urban Movements and Urban Theory in the Late-20th-Century City

MARGIT MAYER

Centered around resistance to urban renewal and the uneven distribution of resources and power and linked to "rising expectations" and political openings, the urban movements of the 1970s and 1980s were part of a broader social mobilization in the aftermath of the various 1960s movements (Castells, 1983). Thus, they often are perceived as coherent and unified actors. Contemporary urban movements, by contrast, are far more heterogeneous, fragmented, and even polarized, and they increasingly play contradictory roles. This chapter explains this transformation and differentiation by relating urban movements to the context of recent urban restructuring and the transformation of the local state. By focusing on two rather similar advanced Western countries, Germany and the United States, my argument assumes that larger restructuring processes and political reorientations have generated structurally new conflict lines in which today's movements are engaged.

I first discuss these new trends and conditions as manifest in U.S. and German cities, highlighting the new local economic interventionism, the restructuring of the local welfare state, and the opening of the local political system that has occurred in both of these policy areas. I then look at the contemporary urban movement sector in each country. These sectors have become extremely differentiated but show more and more similarities across the two countries. Urban movement activities now include community development and housing provision, lobbying and protest around issues of new poverty and marginalization, and direct action and organizing around environmental justice, the co-production of social services, and employment and training programs. The movements' practices have been affected by new municipal, state, and national programs and by changing funding structures that have evolved similarly in the two countries. A third section in this chapter draws out the political implications and strategic consequences for urban movements at the end of the 20th century. Thus, the chapter views recent urban restructuring as a "political opportunity structure" for interpreting the dynamic of urban movements and for exploring and evaluating the latter's shaping of the concrete outcomes of restructuring cities.

The Restructuring of Urban Politics

The Fordist growth model reached its limits around the mid-1970s. This growth model had provided a temporary and prosperous compromise between labor and capital based on mass production, mass consumption, and centrally organized welfare state measures and regional development programs. Within it, municipal politics focused primarily on expanding urban infrastructure and managing large-scale urban renewal. The rigidities of the production structure and the rising costs and destructive side effects of mass production and mass consumption, as well as the politicization of those costs and effects, slowed growth rates and triggered social conflicts and movements (Hirsch & Roth, 1986). With growth declining and loyalties dwindling, the technical and social limits of this growth model became apparent, and Fordist modes of regulation became dysfunctional. Thus, a search was launched, not just to adjust the structure of accumulation (with new, more flexible forms of production) but also for new institutional arrangements and modes of regulation (Amin, 1994). The effects of this on cities can now be read in a growing body of literature on "dual cities" (i.e., on the growing polarization of the economy around a high-paying corporate service sector and low-paying sectors of downgraded manufacturing and lower level services [Brown & Crompton, 1994; Mollenkopf & Castells, 1991]) and in the literature on flexible specialization and the resultant hierarchical differentiation among cities and a new quality of intraurban competition (Krätke, 1991, 1995). These transformations have affected urban politics, producing new trends at the level of local politics regardless of the particularity of the urban regime or the specific history and culture of a place. In Western Europe and North America, local authorities now increasingly engage in economic development and restructure and dismantle the local welfare state while expanding the urban political system to include a variety of nongovernmental actors.

First, increased local economic interventionism manifests itself both in quantitative and in qualitative terms. Quantitatively, local government spending for proactive economic development strategies takes up a growing portion of local budgets. Qualitatively, whereas economic development measures of local authorities once were concerned with the even distribution of (automatic) growth, intervention now is more and more targeted to strengthening endogenous urban and regional development and entrepreneurial initiative. Cities now "market" themselves (in the global economy) by publicizing the virtues of their respective business climates. They seek to make use of endogenous skills and entrepreneurship and to target subsidies to industries promising growth and innovation but also to megaprojects and big festivals. The primary goal of urban policy now is to initiate and stimulate private capital accumulation. Other policy areas are increasingly becoming integrated with or subordinated to economic development measures (Hall & Hubbard, 1998).

An important aspect of this shift in the approach of intervention is that more and more non-state actors have become involved in this local organization of economic conditions. Local authorities support or even establish new institutions for economic development and technology transfer, "roundtables" have emerged locally and regionally to influence policy formation, and other new forms of cooperation are initiated and organized. Depending on the policy area, different actors are involved. For example, labor market policy involves the employment office, social welfare associations, churches, firms, unions, and consultancies as well as the local authorities. In a growing number of policy areas, non-state actors also include the so-called "third" or alternative sector.

The new approach to local economic intervention has changed the formal political structures. To identify the intersecting areas of interest of different actors, a more cooperative style of politics than the traditional top-down approach is necessary. Thus, more pluralistic bargaining systems are being tried. This horizontal style of politics does not mean that these bargaining systems and project-specific partnerships are more open to democratic influence or are more accountable to local social or environmental needs. Participants may, in fact, form rather exclusive groups representing only selected interests.[1] Often, the balance of power is tilted toward business and against unions and environmental and community groups; frequently, one even finds a cleavage between established community groups and more marginalized interests. We also find instances in which traditionally excluded groups participate. The important point is that bargaining and decision-making processes increasingly take place *outside of* the traditional structures of municipal politics. That is why we now speak of local "governance" instead of "government"; the local state has expanded to explicitly include and coordinate a variety of functional interests.

A second trend identifiable in the changing urban politics is related to the first trend. As cities have emphasized economic development, this has redirected resources from other policy areas such as social policy. It also has changed the approach and direction of social policies and has led to a restructuring in the provision of social services. With the subordination of social policies to economic priorities, not only has local government spending for social consumption declined as a proportion of overall expenditure, but a qualitative shift is observable: The traditional redistributive policies of the welfare state have been supplemented or displaced by employment and labor market policies designed to promote labor market flexibility, whereas traditional welfare benefits increasingly are tied to stricter eligibility criteria.

The increase in unemployment, underemployment, casualization of labor, and new poverty cannot be handled with traditional welfare state policies designed to treat such problems as transitory phenomena in a basically full-employment society. The dualization of labor markets, the expansion of precarious and informal jobs, and the shift in social policies have produced a new marginality, the most visible manifestation of which are the tens of thousands of homeless who inhabit major cities. Other, less visible forms of social exclusion and new poverty also concentrate in urban areas, even if their causes are increasingly identified in global processes (Dangschat, 1995; Huster, 1997). Whereas in Germany the lower end of the labor market is, to a far larger extent, unemployed but not as poor as its U.S. equivalent, in the United States it is employed but comparatively poorer. Because central governments have reduced their subsidies, the effects of new risks in the labor market confront local authorities in both countries with a new challenge. Thus, local authorities have had to explore alternative and innovative ways in which to keep their cities functioning. Mostly, they have done so by exploring alternatives in job creation and "workfare" that involve local organizations in both the private and voluntary sectors (Evers & Olk, 1996).

The shift in social policies has yet another element related to economic competitiveness, the first trend. Because the image of cities now plays such an important role in attracting global investment, stern anti-homeless and anti-squatter policies have been drafted, and regular sweeps are now carried out at the showcase plazas of all major cities. This regulation of public space has been observed since the early 1990s, even in cities with progressive governments that have adopted laws prohibiting people from sitting or lying on sidewalks in business districts, just like more "neoliberal" cities (Egan, 1993). To drive beggars, homeless people, and/or "squeegee kids" from the centers of the cities, where they concentrate for a variety of reasons,[2] these groups are being constructed as "dangerous classes" (Ruddick, 1994). Social policies have been abandoned in favor of punitive and repressive treatments of a growing segment of marginalized populations.

The qualitative shift in the orientation of social policies consists of both a restriction of funds and services to the traditional "welfare" clientele *and* a shift toward more active labor market policies, where municipal employment and training programs have been established and where job-creative activities of third sector groups are being supported. These new policies have blurred the traditional distinction between economic and social policies and have created a real link between the local economy and the local operation of the welfare state.

As with the first trend (i.e., the mobilization of local politics for economic development), the restructuring of the local welfare state, the second trend, also involves an opening of governance structures. More non-state actors have become involved in the provision and management of services that once were public-sector led. In the process, these services have been transformed into more active, so-called empowering forms of community (economic) development. This process has made local government into one part of a broader system of service providers. In this expanded system of local politics, the public sector reduces its functions yet plays a more activist role vis-à-vis non-state actors. With less rigid bureaucratic forms and more competition, the local welfare state has become more flexible. Thus, the role of the municipality has changed from being the (more or less) redistributive local "arm" of the welfare state to acting as a catalyst of processes of innovation and cooperation that it seeks to steer, more or less forcefully, in the direction of improving the city's economic and social well-being.

These strategies are being pursued not only in different national and regional settings but also by proponents of different political leanings. No matter whether more progressive forces or more conservative forces dominate a city government, priority is given to economic development policies via the entrepreneurial mobilization of endogenous potential, thereby pushing one of the formerly central functions of the local state—the provision of collective consumption goods and welfare services—into the background. In both cases, more and more public functions are privatized. In both cases, increased engagement in the arena of economic development as well as the provision of social services occurs via new forms of negotiation and implementation involving non-state actors and intermediary organizations. The conservatives are drawn to this model because it involves voluntary action and workfare, allowing state shrinkage. The left finds it attractive because it "enables" people to exercise power for themselves. The liberals pursue it because it emphasizes local community action.

According to these different political/ideological interests, the programs do take on different nuances, and the new bargaining structures differ in terms of their inclusiveness and responsiveness with regard to the interests of neighborhoods, tenants, environmental, or other social movement groups. Depending on the prevailing national political culture, we also have a variety of models of these new partnerships and cooperation arrangements. On one end of the spectrum are those that are strongly entrepreneurial and framed by the rhetoric of a high-level volunteer summit (as in the United States), and on the other end are those that still are more state oriented (as in Germany). In any case, the new bargaining structures are not more biased toward private business than was the old form of urban governance that supposedly emphasized the separation between public and private profit. Nor is it the case that they necessarily prefigure political empowerment within localities, as Amin and Thrift (1994) insinuate. Rather, we are dealing with the state as a contradictory consolidation of antagonistic interests. The concrete form taken by the new institutional arrangements and the degree of their responsiveness and openness for social and environmental interests depend on how actors at the local level seize and struggle over the opportunities created.

■ Urban Movements During the 1990s: Fragmentation and Similarities

The first massive phase of urban movements at the end of the 1960s and during the 1970s occurred when U.S. neighborhoods exploded in a "community revolution" and German citizens initiatives staged large-scale mobilizations against renewal projects and in defense of residents' living conditions. These mobilizations soon expanded into broad struggles over the cost and use value of public infrastructure into which leftist projects of the gradually dissolving New Student Left frequently inserted themselves. The subsequent phases, however, saw the disintegration of these broad coalitions (for Germany: Mayer, 1993; for the United States: Fisher & Kling, 1993). A transition from "protest to politics" as well as to service delivery set in by the late 1970s in the United States and later in Germany, but by the end of the 1980s, most of the housing and self-help initiatives as well as groups formed around a variety of local social and economic problems had entered into a different structural relationship with the local state. Their formerly antagonistic relationship gave way to a stance of working "within and against" the state (Mayer & Katz, 1985). Meanwhile, these movements have become essential constituent agents in the co-production, development, and implementation of social and cultural services, housing provision, and local economic development, thereby capturing access to and some of the wealth of the city while helping to process new types of conflicts and exclusions. The dynamic of this new work in cooperation with, and often funded by, the local state sets these types of movement groups apart from others that continue to mobilize outside such a relation with the state.

I distinguish movements around new forms of urban development, which include campaigns and mobilizations against specific development projects but also movements in defense of threatened communities, from movements of the newly marginalized, so-called poor people's movements that in major cities have garnered resources and found stages and sufficiently supportive publics for disruptive and sometimes successful collective action. These movements correspond to the novel trends that recent urban research has identified as significant in the field of local politics—the expansion of the urban political system, the new competitive forms of urban development, and the erosion of welfare rights.

Routinized Cooperation With the Local State

The opening up of the urban governmental system to nongovernmental stakeholders, and the strategy of many municipalities to employ former social movement organizations in the development and implementation of (alternative) social services, cultural projects, housing, and economic development has been an important force shaping the trajectory of urban movements since the 1980s. This shift in the political opportunity structure helps to explain the transformation of the self-organized initiatives around housing and the movement groups active around social and economic problems that became manifest in many cities during the 1970s and 1980s. By including and funding third sector organizations, municipalities have sought to achieve political vitalization as well as financial relief, although these goals frequently conflict with each other.

This development is rather more advanced in North America than in Germany. A particularly germane illustration is provided by an organization called Banana Kelly in the Bronx, New York. This community organization emerged from squatting and militantly defending houses during the 1970s, went on to rehabilitating, and is now managing hundreds of low-income homes, helped along by a variety of municipal programs funding "sweat equity" and tenant management. It now is active not only in housing but also in economic development. It has even gone global in search of an investor and found a Swedish firm to set up a large paper mill for recycling Manhattan's enormous output of office paper. Besides job creation, Banana Kelly also engages in education and training programs including bringing Los Angeles gang members to Brooklyn to teach them family values and community respect. In 1996, the group received a Best Practice Award at the Habitat Conference (Harris, 1995; Holusha, 1994; Rivera & Hall, 1996).

Although Banana Kelly obviously is a bigger and more "successful" case than many, it demonstrates the polyvalent functions that community-based and client-oriented groups have begun to play in cities all over North America and Western Europe (for North America: Fishman & Phillips, 1993; Rich, 1995; Shragge, 1997; von Hoffmann, 1997; for Germany: Froessler, Lang, Selle, & Staubach, 1994; Rucht, Blattert, & Rink, 1997; Stiftung Mitarbeit, 1995; for comparative studies: Selle, 1991). The establishment of alternative renewal agents and sweat equity programs and the funding of self-help and social service groups have been long and contested processes, and since the late 1980s, municipal social and employment programs everywhere have been making use of the skills, knowledge, and labor of such movement groups. Similarly, many cultural projects have become part of the "official city," and many youth and social centers play acknowledged roles in integrating "problem groups" and mitigating potential conflict.

When first launched in the United States during the early 1970s, these programs were far from coordinated, far-sighted adaptations of regulation mechanisms. Rather, they were disparate and uncoordinated reactions to the pressure of tenant groups and community organizations, on the one hand, and to the financial crisis of cities, the renewal problems of decaying neighborhoods, and the threat to social integration posed by minority and poverty populations, on the other. For example, in New York City, the Community Management Program (launched in 1972) and the Sweat Equity Homesteading Program were not coordinated until 1978 in one administrative unit, the Division of Alternative Management Programs (DAMP) within the Office for Urban Renewal. All the different DAMP programs required self-labor (i.e., sweat equity) of the tenants. Soon, the municipal subsidy for this type of self-help and self-organization proved "successful" for the city as rent payments went up and the rate of privatization was accelerated (Mayer & Katz, 1985). As more and more municipalities began to contract out services and training programs, movement groups began to seize these opportunities to make themselves indispensable.

On the national level, the administration of President Jimmy Carter made neighborhood organizations a central component in the "partnership to build cities" in 1978, and its Office of Neighborhood Development distributed funds that subsidized program development and administration and staff salaries for community organizations. These concessions to neighborhood-based groups and movements were modest in comparison to the billions of dollars authorized under the Model Cities and Community Action Programs of the 1960s, but they were designed to support privatization processes through state instruments and to systematically include the private sector in urban revitalization.

With President Ronald Reagan's "new volunteerism," the use of societal self-organization reached a qualitatively new level. Many federal programs were cut back or eliminated, and the remaining funding was targeted for specified projects. In response, movement groups gradually transformed themselves into co-producers/administrators of public goods and services. A variety of pilot programs manifested the search for viable tripartite arrangements, exploring what role the state might play in restructuring certain labor markets and modes of production; model Enterprise Zone Programs as well as the Alternatives to Service Delivery required the existence and participation of community-based organizations. A 1982 program, Partnerships for Service Delivery, called on neighborhood organizations to develop "creative and innovative arrangements" for delivering and organizing services in a variety of municipal policy fields such as the environment, crime, health, and education. Another program bestowed awards, so-called Community Development Partnerships, on neighborhood organizations that succeeded in matching private funds to Community Development Block Grant (CDBG) funds.

The Reagan administration continued to support the National Center for Neighborhood Enterprise, which tested and propagated the capacity of neighborhood groups to become entrepreneurial. Other demonstration projects were launched under the name of Quality of Life Initiatives by the Department of Housing and Urban Development; for example, the National Self-Sufficiency Project provided funds to facilitate the move from "welfare dependency" to productive employment for "highly motivated" single mothers. The Minority Youth Training Initiative in 1983 combined training of young people in housing rehabilitation and management with job placements using a partnership of mayors, public housing agencies, and the private sector. All of these programs supported and subsidized active, community-based interest organizations. They made community groups a required partner in bargaining structures, allowed them to function as conduits for community development funds, but also forced them to adjust to the economic norms of the public-private partnerships (Mayer, 1987b). During this period, nationwide intermediary organizations, such as the National Civic League, Neighborhood Reinvestment Corporation, Local Initiatives Support Corporation, Enterprise Foundation, and National Community Development Association, emerged in support of local groups. Thus, the ground was prepared for the early 1990s, when both the Democratic White House and the Republican "Contract With America" advocated "community empowerment" strategies as a way in which to tackle the crisis of the cities (Boyte, Sirianni, Barber, & Delli Priscoli, 1994; Dreier, 1993). President Bill Clinton's National Urban Policy Report of 1995 specified the goal: "[to] promote solutions that are locally crafted and implemented by entrepreneurial public entities, private actors, and the growing network of community-based corporations and organizations." The Clinton administration further strengthened the role of community-based organizations through the Empowerment Zones/Enterprise Communities program in 1993—$3.5 billion for 10 years—by requiring partnership and cooperation of all relevant local stakeholders. Clinton's national urban policy further supported a federal network of community development banks and microenterprise programs with subsidies from the Small Business Administration.

At the same time, foundations shifted their funding criteria. After having supported brick-and-mortar projects for 20 years, during the late 1980s, foundations began to emphasize the formulation and implementation of comprehensive revitalization plans. Thus, a wide variety of locally based movements that had their origins in various local struggles and differing political traditions developed their practices in line with this history of national and local programs as well as in response to foundations' policy changes.

Germany does not have as thickly developed a community-based infrastructure as does the United States. Movements have been less territorially based, and state programs reacted to the specific German political and institutional culture. Neither the old Federal Republic of Germany nor the German Democratic Republic had a tradition of strong civic voluntarism supported and bounded in by the state. Not until the new self-help movements of the 1980s did a civic activism rhetoric emerge ("subsidiarity"), which in a few states has led to programs with a neighborhood accent.[3]

German municipalities first began to launch such programs in response to the squatters' movement and alternative movements active around women's issues, immigration, drugs, and other social concerns. Distinct programs were geared to housing and urban renewal, self-help and social policies, and a third set of job creation programs targeted to "groups with problems in the labor market."

The first such step was a self-help rehabilitation program launched by the Berlin Housing Senate that included various intermediary organizations and both technical assistance and socially oriented renewal agents in the planning, formulation, and implementation of housing and renewal policies (Boll, Mauthe, & Osoriao, 1991; Clarke & Mayer, 1986, p. 412; Mayer, 1987a; Schubert, 1990).[4] Distinct from this, the Christian Democratic Union (CDU)-dominated Senate also launched a social services program in 1983 in response to the demands of alternative collectives and citizen initiatives. Although its umbrella organization, AK Staatsknete, had demanded a self-administered fund from various departments, the CDU offer was restricted to social services and health-related activities but geared toward projects based on client self-help and voluntary co-production of health services (Fink, 1983; Grottian, Krotz, Lütke, & Pfarr, 1988). In Munich, a similar funding program was established by a Social Democratic Party city government in 1984. It was called the "Munich concept for supporting self-help groups and self-organized projects in health and social services." Like the Berlin program, it complemented the existing system of social service provision with activities emphasizing self-organization and voluntarism.[5] The municipal employment programs of this early phase were more properly social programs. They targeted so-called problem groups in the labor market, subsidizing their unemployment or welfare benefits to make some type of employment possible, often in the irregular "second labor market." Here, the city of Hamburg spearheaded the innovation with its Second Labor Market Program established in 1982 (Fiedler & Schrödter, 1983). Other cities and some states followed with funding programs for job creation, centers for the unemployed, and technical assistance for project management.[6]

As in the United States, diverse funding sources eventually were combined—in the German case, funds from the labor office, through social assistance schemes, European Community/European Union and state programs, youth services, and the like; in the U.S. case, CDBG moneys, funds from local and state programs, foundation support, and cooperative arrangements between different offices within and outside the municipality. Movement groups turned into nonprofits and engaged in implementing the programs as partners of local government as well as of welfare associations/charities, unions, chambers, and the unemployment office.[7] Although the programs initially were organized along different policy sectors—in urban renewal and housing rehabilitation, in social, or in women's projects—and although only a relatively small number were explicitly concerned with economic reproduction and labor market problems, the emphasis has since shifted, and the integration of different policy fields has become the norm. This shift was triggered by the cuts in urban infrastructure and social services that set in after 1978 in the United States and reached a climax with Clinton's welfare reform of 1996. In Germany, old funding programs and federal job creation funds have been reduced more recently. Groups now scramble for various European Union funds and state and local programs that target labor market integration and presuppose the inclusion of community actors in the endogenous development of localities.

In the United States, the emphasis of poverty alleviation has shifted from welfare to workfare programs; economic development and housing programs have shifted from brick-and-mortar approaches to microenterprise programs integrating both social welfare and economic development goals (Fishman & Phillips, 1993; Kingsley, McNeely, & Gibson, 1997; Servon, 1997). In Germany, similar efforts toward comprehensive, integrated community development are underway but face more obstacles. The temporary, public sector, job creation schemes of the Federal Labor Office (so-called *Arbeitsbeschaffungsmaßnahmen*) have made these programs into social programs rather than into policies encouraging market success. This particular German history contributes to the unique difficulties in "entrepreneurializing" social labor within the German labor market. Contemporary efforts to make work requirements flexible for welfare recipients face much bigger barriers in Germany than in the United States, where workfare (including the right to earn and keep a wage or to start a small business) has long been a part of social policies.

The rigid German laws regulating social and labor market policies recently have begun to be adjusted. Starting in 1992, municipalities have launched "*Arbeit statt Sozialhilfe*" ("work instead of welfare") programs, and in 1994 the federal government complemented the existing welfare law with a *Hilfe zur Arbeit* aspect, a workfare component that allows nonprofit organizations and communal businesses to employ and train welfare recipients in housing renewal, solar technology, or community restaurants.

In spite of such national differences, the various programs tying movement groups into social services, urban renewal, and job creation initiatives gradually blurred the boundaries among these different policy areas. Social and housing and alternative projects have had to expand their job creation and training capacities. Movement groups working with these programs have learned to combine strategies tackling social marginalization or urban repair with job creation measures and have begun to relate their work to the problems of the labor market or to groups structurally disadvantaged in the labor market. In the process, and in both countries, a new approach built around collaborative and comprehensive strategies has emerged. Groups that integrate social, ecological, housing, and job creation goals in an entrepreneurial fashion now thrive.

The bulk of the research focusing on these novel forms of institutionalization of social movements within the shifting relations of welfare systems and provision emphasizes the "contestatory character of their constituency" and the counterweight they pose to "conventional views of local economic planning" (Lustiger-Thaler & Maheu, 1995, pp. 162, 165). Whether in the economic development sector, alternative services, or women's projects, the work of the groups generally is found to be an innovative and progressive challenge to public policy and a contribution to improving access to the local political system and to providing potentially more active citizenship (Clarke, 1994, pp. 9-10; Froessler et al., 1994; Hopkins, 1995; Jacobs, 1992).

Closer examination, particularly of the more recent developments, reveals that (former) movement organizations that have inserted themselves into the various municipal or foundation-sponsored funding programs play a rather complicated role within the urban movement scene. On the one hand, they enhance organization building and lend stability to the urban movement infrastructure and, therefore, to the conditions for continuing mobilization. On the other hand, the widening and growing internal differentiation within the movement sector has led to new conflicts and antagonisms. Movement organizations now participating in the new governance arrangements are subject to the danger of institutional integration—"NGOization" or the process of being turned into a nongovernmental organization (NGO) (Demirovic, 1998)—and the pursuit of "insider interests." Their own democratic substance is far from guaranteed (Fehse, 1995; Lang, 1995; Roth, 1994). As these organizations find themselves threatened by cuts and faced with the reorientation of public sector programs toward labor market flexibility, competition among them for funding intensifies, and the groups engage more in private lobbying strategies to secure jobs and finances than in creating public pressure. Furthermore, the alternative renewal agents and community-based development organizations that are busy developing low-income housing or training and employment opportunities for underprivileged groups find themselves criticized and attacked by other movement actors who do not qualify for inclusion or who prefer squatting or other nonconventional forms of action.

Opposing the New Competitive Urban Politics

As urban leaders upgrade their localities in the international competition for investors, advanced services, and megaprojects; as they rebuild their downtowns into producer-oriented service centers triggering gentrification, displacement, congestion, and pollution; and as neighborhoods not fitting into this design are abandoned or turned into sites for unattractive functions, opposition movements have formed. They have either built on existing networks or sprung up anew. They range from defensive and pragmatic efforts to save the prevailing quality of life or privileges (which sometimes are progressive, environmentally conscious, and inclusive, but other times are selfish, anti-immigrant, and racist) to highly politicized and militant struggles over whose city it is supposed to be (as in anti-gentrification struggles or movements against other growth policies).[8]

Recent social movement research has emphasized mobilizations seeking to protect the home environment from traffic, development, or projects that people do not want to have "in their own backyard." Such groups quickly become skilled at a variety of tactics and repertoires such as petition drives, political lobbying, street confrontations, and legal proceedings. For many of them, social justice, which characterized the goals and practice of citizens initiatives during the 1970s, has been replaced by particularist interests and/or a defense of privileged conditions[9] (Krämer-Badoni, 1990; Krämer-Badoni & Söffler, 1994). Other local movements are made up of working class and middle class participants mobilized against highway construction plans, traffic congestion, or housing shortages.[10] Particularly in the United States, minority/working class communities have mobilized against the polluting industries and hazardous facilities with which they are disproportionately burdened. The action repertoire of such groups goes well beyond that of the defensive "not in my backyard" movements. In addition to direct action (e.g., demonstrations, blockades, corporate campaigns) to put public pressure on polluting firms, they also undertake independent analyses of urban problems and demand representation on relevant decision-making boards (Bullard, 1993; Di Chiro, 1992; Russell, 1990).

An American study that surveyed 325 local groups "run by the poor for the purpose of changing social structures" (McCarthy & Castelli, 1994, p. 11) calls them "poor empowerment groups." Corresponding to the more pronounced sociospatial differentiation in U.S. cities, these movements organized on the basis of locally manifest discriminations. Their achievements range from pressuring banks and insurance companies to reinvest in low-income communities, to improving local working conditions, to committing the city to rehabilitate abandoned buildings and to invest in neighborhood development rather than just downtown development.

Frequently, movements against urban growth policies and gentrification are directly triggered by the new development instruments of big city politics. These include large, spectacular urban development projects such as Berlin's *Potsdamer Platz;* festivals such as the Olympics, World Expo, international garden shows, and 1,000-year birthdays; and the attraction of mega-events, sports entertainment complexes, and theme-enhanced urban entertainment centers, all of which depend on the packaging and sale of urban place images (Häußermann & Siebel, 1993; Scholz, 1997). Movements attack the detrimental side effects of, and the lack of democratic participation inherent in, these strategies of restructuring the city and of raising funds. They criticize the spatial and temporal concentration of such development projects and complain that the concentration on prestige projects detracts attention and finances from other urban problems and restricts investments in other areas. Protest campaigns against the instruments of city marketing raise questions of democratic planning that urban elites concerned with intraregional and international competitiveness would rather downplay.[11] Sometimes, these opposition campaigns bring otherwise scattered local movement groups together in broad coalitions (as happened, e.g., in the NOlympia Campaign in Berlin during 1991-1993). Radical, so-called autonomous movements take the lead by consciously seizing on the importance that image politics has gained in the global competition of cities and by devising image-damaging actions to make their cities less attractive to big investors and speculators.

Against the Erosion of the Welfare State

In reaction to the dualization of the labor market and the shift in social policies, new poor people's movements have sprung up, often accompanied by actions of supporter groups and advocacy organizations as well as anti-racist initiatives.[12] Most social movement researchers assume, erroneously, that this population is not only poor and without resources but also disempowered and passive. In fact, the resources of these groups consist primarily of their bodies and time, so their protest activities tend to be episodic and spontaneous, local in nature, and disruptive in strategy.[13] At best, their disruptive tactics block normal city government operations and threaten the legitimacy of local policies of exclusion. This was the case when the homeless in Paris defended their right to the city in a campaign around the slogan "*droit au logement*" ("the right to housing"). Their actions culminated in a spectacular squat in December 1994 in the middle of the sixth arrondissment a few months before the presidential elections (Body-Gendrot, in press; Péchu, 1996). In general, however, such new poor people's movements face an increasingly recalcitrant and punitive state. Only under rare conditions does setting up encampments, holding public forums, and making demands on the city for purposes of resisting efforts to drive them out of the downtowns allow them to develop solidarity, political consciousness, and organizational infrastructures, precisely the elements that social movement research assumes as preconditions for the emergence of mobilization. David Wagner describes such conditions in his study of homeless demonstrators in Portland, Maine. Wagner (1993) observes how the 100 live-in participants of a "tent city" were empowered by their action and were able to achieve significant concessions from the city. Other localized tent cities and other struggles by newly marginalized groups have been less successful, for example, the battles over New York City's Tompkins Square Park, where housing and anti-gentrification activists joined the homeless and squatters who had erected about 100 structures in the park at the peak of the movement.[14]

Where resource-rich, political advocacy groups dedicate themselves to the problems of the homeless, or where professional activists make their resources available to such organizations, durable and effective mobilizations can be achieved (Péchu, 1996). This was the case with the Paris groups Comité des mal-logés (CML), which was founded in 1987 and had 1,300 members by 1990, and Association Droit au Logement, which practically replaced CML after 1990 and unites about 8,000 people in its broader milieu. Cress and Snow (1996) also found, in their ethnographic study of 15 homeless initiatives in eight U.S. cities, that 75% of resources come from "outside," which in the United States means charity, church, and civil rights support organizations rather than political activist organizations (see also Wright, 1997). Organizations sponsored by the Industrial Areas Foundation and similar forms of U.S. community organizing in the Saul Alinsky tradition have for many years relied on the networks of churches. Such church-based efforts enjoyed a significant revival in poor neighborhoods during the 1990s (Cortés, 1996; Harvey, 1998).

In addition to churches, political activist groups, and local coalitions (Bartelheimer & von Freyberg, 1997; Blum, 1996; Mette & Steinkamp, 1997; Stiftung Mitarbeit, 1995), another relevant support network in Germany is provided by autonomous movements and anti-racist initiatives. These latter groups publicly attack the production of new poverty, homelessness, and xenophobia while also mobilizing against their own eviction from squatted buildings and "liberated areas" of city centers. Anti-racist initiatives have formed lately because of the police raids carried out to "clean" the citadel plazas of immigrants and the poor. In many cities, so-called danger zones have been marked from where individuals looking "suspect" may be deported[15] and against which newly formed initiatives stage protest demonstrations, provide legal aid, and put public pressure on the local governments. In June 1997, a "Downtown Action Week" took place in 19 German and Swiss cities to create public awareness and pressure around the widespread practice of driving out the new marginality from core areas. In June 1998, the same coordinated campaign focused on train stations, a newly contested terrain between marginalized groups and the security and consumption interests of more hegemonic groups.

Another new movement form has emerged in the context of housing need and new poverty, although its members do not see themselves as a "poor people's movement." The majority of the so-called *Wagenburgen* (groups of people squatting on vacant land, living in trailers, circus wagons, or other mobile structures) see their actions as "a form of resistance against the political, social, and economic relations in this city and this country" (*Vogelfrei,* 1994). About 70 to 80 such sites exist in Germany (Knorr-Siedow & Willmer, 1994). Of the 15 in Berlin, most have been in the downtown area and, therefore, been threatened by eviction or already displaced to other locations. Their political orientations cover a wide spectrum; whereas some use the freedom that this lifestyle allows them for political activism (e.g., sheltering refugees without legal status), others are content to explore alternative ways of living. Evictions and the threat of evictions have brought them together in campaigns to pressure city governments to tolerate the sites, delay construction, and/or provide other acceptable locations.[16]

Current conditions in the labor market and the shift from social welfare to more punitive workfare policies have influenced the urban movement scene in other ways. Not only have hundreds of new organizations sprung up, but the number and variety of institutions and projects "servicing" the marginalized have exploded as well. What connects them to the first category of movements is their functioning within municipal programs that harness the reform energy of community-based groups. Going further than the community-based organizations described in the first group, they seek not just to "mend" the disintegration processes that traditional state activities cannot address but also to develop innovative strategies that explicitly acknowledge the new divisions within the city. An example would be a grassroots organization called Proyecto Esperanza in Los Angeles. It helps recent immigrants to find jobs and places in which to live by training them to find work in the growing informal sector as day laborers rather than by channeling them into normal job training programs (Hopkins, 1995, pp. 41, 56). German organizations similarly seek to connect the contemporary labor market reality with social goals, for example, LIST/Zukunft Bauen in Berlin, which uses second labor market programs to train unemployed youths as cooks or in housing rehabilitation (Zukunft Bauen et al., 1994, p. 12). Although of later and different origin, these new poor people's movements are becoming part of the routine cooperation with the local state, just as the community groups and self-help initiatives did during the period of high progressive mobilization. Consequently, they face the same types of dilemmas.

Under certain conditions, such projects manage to challenge the state while simultaneously exploiting its workfare program for their own goals of creating solidarity and empowerment. A case in point would be Chic Resto-Pop, a Montreal community restaurant/nonprofit organization started in 1984 by 12 welfare recipients. It provides jobs for the poor in the community and offers inexpensive meals (for 800 people). Its trainees (currently 93) participate in a workfare program, but the organization also is mobilizing locally and demanding the transformation of this very workfare program, arguing both for local control and government support for the emerging social economy (Shragge & Fontan, 1996, p. 8; for other cases, see Shragge, 1997). More frequently, projects are not aware that official politics increasingly looks to NGOs and community groups to replace state politics and to function as repair networks for economic and political disintegration.

Urban Theory and the Perspective of Urban Movements

The specific sociospatial context that different cities provide for contemporary social movements and its consequences for the dynamic and development of urban movements need to be further differentiated. The new international division of labor and new global urban hierarchies have generated urbanization patterns fundamentally different and more differentiated from those of the Fordist era. Recent urban research has analyzed the emerging structural differences within the urban system, explaining them with a variety of approaches—as a process of economic-functional hierarchization (Krätke, 1995, p. 126), as uneven clustering of different roles and functions within a transnational system (Logan & Molotch, 1987, p. 258), as polarization between the organizing nodes of the global economy and subordinate cities within a global urban network (e.g., the global city literature). Correspondingly, the patterns of conflict and movements in the different types of cities will vary. The homogeneous pattern of conflict characteristic of the Fordist era is dissolving. Cities at the top of the global hierarchy develop particularly pronounced conflict patterns reflecting the internationalization of their working classes and neighborhoods, their precarious labor relations, and their eroding municipal powers. At the same time, the mere size of such metropolitan regions facilitates the emergence of a critical mass of disaffected people, a precondition for the building of movement milieus and the construction of collective projects and identities. This is where movements against central city development servicing global headquarters as well as new poor people's and anti-xenophobic movements proliferate and may expect support.

Old deindustrializing cities, on the other hand, feature struggles over plant closures and new employment creation and, depending on the profile with which the cities attempt to reposition themselves in the new urban hierarchy, more or less intensive cooperation between the municipalities and community groups. Cities trying to make their fortunes as "innovation centers" (Logan & Molotch, 1987, pp. 267 ff.) frequently provoke environmental protest and slow growth movements. Even though these and other differences among urban movement milieus are far from adequately researched, our overview of the changes that urban movements have undergone and of their characteristics during the current period allows some preliminary conclusions about their role and future possibilities for action.

My overview brought to light a variety of movements active in the urban terrain. Occasionally, their social bases overlap but just as often are distinct in terms of organizational patterns, goals, and tactics. Within each city, the interaction and coalitions among these different strands is of crucial importance for the impact that social actors have on the shape of the city. Here lies a weakness of the otherwise excellent and useful analyses of recent urban restructuring: When turning their attention to conflict and political action, most authors capture and generalize from just one type of urban movement activity. Whereas Saskia Sassen points to the "presence" and increase in representative power that the (immigrant) marginalized have achieved in global cities (Sassen, 1997), David Harvey's current work highlights the Living Wage Campaign carried out by groups such as Baltimore's BUILD (Harvey, 1998). Both are optimistic about the role of these movement actors. Mike Davis, on the other hand, paints a bleaker picture by emphasizing the mobilization of homeowners in defense of home values and neighborhood exclusivity (Davis, 1990). For Lustiger-Thaler and Maheu (1995), urban social movements *are* community economic development corporations; for Neil Smith, they *are* the movements of the marginalized and the homeless (Smith, 1996). Misleading conclusions are drawn from whichever fragment of today's complex movement scene "fits" the respective screen.

Urban theorists, then, have a hard time opening up to the complex and heterogeneous urban movement scene of the 1990s. Social movement theorists more typically make the environmental, women's, peace, and human rights movements—rather than the urban movements—their "objects" of study and of generalization. To the extent that local/urban movements are analyzed, the interest is more in their unique local character than in distilling particular patterns and roles that might be generalizable across cities and perhaps even across countries.

We need to draw on both urban theory and social movement theory for analyzing the broad spectrum of movements active in today's cities and for assessing their effects. If we adapt the findings of advanced contemporary urban analysis as a "political opportunity structure" for interpreting the dynamic of today's urban movements, then we will be better able to understand the effects that the expansion of the urban political system (i.e., the shift from government to governance), new competitive forms of urban development, and the erosion of traditional welfare rights have had on urban movements. This goes beyond assessing the changing *level* of activity of urban movements as, for example, Pickvance (1995) has undertaken. Quantitative shifts in the occurrence of urban social movements might well be assessed with a large cross-national framework. Explaining such transformations requires a more empirically grounded analysis, which is why this chapter pursues a more limited comparative scope. Focusing on a "similar countries comparison" allows us to attend to both movement characteristics and the role of contextual features, and it enables us to understand contextual features as more than merely a list of contingent variables. Within the historically specific contexts of two advanced Western countries, contemporary urban movements are playing a contradictory role in contributing to and challenging the shape and regulation of the city. Although their practice with innovative urban repair and their inclusion in municipal governance structures might well feed into the search for (locally adequate) post-Fordist solutions and arrangements (making the movements appear "functional" and co-optable), their challenge to undemocratic and unecological urban development schemes might yet contribute to a more participatory and more sustainable First World model of the city. To realize this potential, however, the new problems confronting contemporary urban movements must be addressed.

One of these new problems is the new antagonisms within the movement sector created by the restructuring of the urban polity that has expanded and includes some, but not others, in its governance arrangements. While lending stability to the movement sector, newly inserted groups also contribute to its fragmentation and polarization. Because competition for funding has intensified, groups engage more in private lobbying strategies to secure jobs and finances for their own constituents rather than creating broad public pressure. Some organizations also have found themselves attacked by other movements. Such tensions were expressed in violent actions by autonomous groups against Stattbau, the alternative renewal agent in Berlin. Rehabilitation of old buildings often prepares the way for gentrifiers, so protests were directed toward the symbols of advancing gentrification, such as chic "yuppie" restaurants, and against the intermediary organizations that were seen as organizing these processes (Kramer, 1988). Similar tensions have been observed between squatters and community development organizations on New York's Lower East Side and have flared up in the struggles around Tompkins Square Park (Smith, 1996, pp. 3-29).

Next to these new antagonisms, a second new problem demands attention: The inclusion of movement groups in revitalization and other partnerships has meant that many have become tied up with managing the housing and employment problems of groups whose exclusion by normal market mechanisms otherwise might threaten the social cohesion of the city. Finally, a third problem arises from pressures to "entrepreneurialize" the social and community work of these groups given that funding support for them is increasingly available only through workfare programs and microenterprise arrangements. These structurally new constellations have to be acknowledged, and their specific constraints (and the opportunities peculiar to them) have to be identified.

New institutional avenues also offer opportunities to tackle new problems. The growing role of local politics (even within global contexts), and the simultaneous dependence of city governments on (former) social movement organizations for processing the complex antagonisms within contemporary cities, enhances the chances for tangible movement input. Whereas this dependence involves routinized cooperation between the local state and former social movement organizations with regard to community economic development, client-based social services, and women's centers, these new partnership relations also are beginning to influence interaction between the local state and various slow-growth and poor people's movements. The eroding local competence of many city governments increases the pressure on local political elites to negotiate and bargain with movement representatives within the channels and intermediary frameworks generated by the routinization of alternative movement labor in the context of municipal (employment or revitalization) programs. Thus, movements that take stands on the use value of the city, such as ecological and poor people's movements, now profit from the new culture and institutions of nonhierarchical bargaining systems, forums, and roundtables. (Obviously, these new structures of governance are open to the less progressive, xenophobic, and antisocial movements as well.)

Thus, the struggle for a democratic, sustainable, and social city crucially requires forging coalitions among different strands of urban movements. That struggle will be successful only if the newly available avenues are not used defensively or to protect threatened individual privileges. Some urban theorists have seen this struggle as one between global elites and local communities and reduce it to the simple antagonism between distant powerful forces (e.g., global capital) and local victims "retrenched in their spaces that they try to control as their last stand against the macro-forces that shape their lives out of their reach" (Castells, 1994, p. 30). Viewing local movements as "innocent and good" vis-à-vis distant forces of domination and power would have been problematic for the 1960s and 1970s, when the majority of urban movements still were part of a larger social struggle against broadening forms of domination. Today's urban movements certainly cannot be seen, in their entirety, as part of emancipatory struggles. They are contradictory and complex agents in the shaping of post-Fordist cities.

We have to recognize the new fragmentations within the movement sector as well as with the massive marginalization and social disintegration increasingly characteristic of urban life. The institutionalized, professionalized, and entrepreneurial movements that now benefit from routinized cooperation with the local state frequently want nothing to do with younger groups of squatters or cultural activists. Because of their preoccupations with the new funding structures, they often are quite distant from the growing marginalized and disadvantaged social groups. Because the latter's organizations and forms of resistance do not automatically lead to mobilization or widespread support, it becomes crucial that those parts of the movement sector that enjoy stability, access, resources, and networks devote part of their struggle to creating a political and social climate in which marginalized groups can become visible and express themselves. Only if these movements interact, politicize the social polarization inherent in the post-Fordist city, and build on the mobilizing potential of the new inequalities will the struggle for socially just, environmentally sustainable, and democratic cities have a chance. Where movement actors acknowledge and make transparent their new dependencies (on both the state and the market), they manage to identify and exploit the opportunities given by current constellations. Where movement actors succeed in transforming the funds and the stability of the resource-rich movements into support for precarious movement groups, different parts of the movement can strengthen each other. Under such conditions, existing opportunities, whether workfare programs or anti-poverty initiatives, can be seized and used to attack and restrict marginalization and discrimination at the root of the new poor people's movements.

NOTES

1. I reject the position that new bargaining systems based on negotiation represent "in short, a better form of democracy" (Amin, 1994, p. 29).

2. This is where the institutions that service them are located and where public space exists that allows for social relations.

3. The state of Northrhine-Westphalia has made neighborhoods "with special renewal needs" an emphasis in a major funding program (Lang, 1994). See also Ministry of Social Affairs (n.d.).

4. Alternative renewal agents called *Stattbau* emerged first in Berlin (in 1982) and Hamburg (in 1984) and were later followed by similar ones in other cities.

5. In both "forerunner" cities, the largest amount of funding went to self-help centers and to contact places established to mediate between grassroots self-help groups, on the

one hand, and the state and welfare bureacracies, on the other. By 1988, such self-help contact centers existed in 20 West German cities through a national program. Another model program was started in 1992 to establish 17 new centers in East German states ("Selbsthilfe ist gefragt," 1992).

6. Most advanced was the program of the state of Northrhine-Westphalia (Matzdorf, 1989).

7. Even in the former German Democratic Republic, the citizens movements of 1989 soon were displaced by social movements that rapidly reached the level of formalization and professionalization that Western groups had reached more slowly. Rucht et al. (1997) show how this adaptation was enforced through funding programs and labor market instruments implemented since 1991.

8. The defensive and pragmatic mobilizations to protect the home environment cover a wide spectrum, from mobilizations against highway construction plans, traffic congestion, or polluting industries to those that mobilize for privatistic and basically exclusive interests. An interesting case of the latter is the mobilization directed against the Berlin government's decision to relocate the former residents of an evicted downtown encampment to the fringe of the Spandau district. Local residents, supported by the Bezirk administration, formed a citizens initiative, staged protest marches, and collected thousands of signatures to prevent the dumping of metropolitan poverty "in their backyard." On the other hand, anti-gentrification struggles have occurred in New York, Paris, Amsterdam, and Berlin, with the most well known being that on the Lower East Side of Manhattan around Tompkins Square Park (Smith, 1996). So-called slow growth or no-growth movements have emerged in the sprawl of Los Angeles as well as in the new peripheries of Washington, D.C. (Sambale, 1994; see also the series, "Growing Pains," appearing in *The Washington Post* during the winter of 1996 and spring of 1997).

9. An example would be actions directed against housing for asylum seekers in Europe or against public housing for minorities in the United States.

10. The quantity of studies is no indicator of the quantity of movements; it merely allows conclusions to be drawn about the research interests of the authors. An impression of the spread and practices of such movements can be gained from journals such as *Everyone's Backyard: The Journal of the Grassroots Movement for Environmental Justice,* published by the Center for Health, Environment, and Justice in Washington, D.C.

11. One function of such megaprojects is supposed to be a socially integrative one, that is, enhancing residents' identification with the city. More often than not, this top-down strategy manipulates residents' interests.

12. The term refers to the study by Piven and Cloward (1977) that demonstrated, using the cases of the unemployed movement of the Great Depression, the civil rights movement of the American South, and the welfare rights movement led by the National Welfare Rights Organization, that protest movements of the poor tend to be more successful when they are spontaneous, disruptive, and radical than when they build on stable national organization and calculable, moderate conflict repertoires. The concept "poor people's movements" emphasizes, in contrast to other approaches in social movement research, the capacities for agency even among resource-weak groups.

13. In North American cities, where the urban conflict structure is shaped by more intense racial conflict and sociospatial polarization than in Europe, these movements also may take the form of riots, as occurred in 1992 in Los Angeles.

14. The taking and defense of the park were organized around slogans such as "Housing is a human right" and "Gentrification is genocide" (Ferguson, 1991, p. 25).

After closure of the park in 1991, the homeless set up new shanties and tent cities on several empty lots in as yet ungentrified neighborhoods east of the park, from which they were repeatedly evicted (Smith, 1996).

15. These municipal actions have been encouraged by the "Action Security Net" passed by the state and federal ministers of the interior to make Germany's cities "safer and cleaner" (*Tageszeitung,* 1998).

16. For Berlin, see Natz (1993).

REFERENCES

Amin, A. (Ed.). (1994). *Post-Fordism: A reader.* Oxford, UK: Basil Blackwell.

Amin, A., & Thrift, N. (1994). Living in the global. In A. Amin & N. Thrift (Eds.), *Globalization, institutions, and regional development in Europe* (pp. 1-22). Oxford, UK: Oxford University Press.

Bartelheimer, P., & von Freyberg, T. (1997). Neue Bündnisse in der Krise der sozialen Stadt: Das Beispiel der Sozialpolitischen Offensive Frankfurt. In W. Hanesch (Ed.), *Überlebt die soziale Stadt? Konzeption, Krise, und Perspektiven kommunaler Sozialstaatlichkeit* (pp. 173-212). Opladen, Germany: Leske & Budrich.

Blum, E. (Ed.). (1996). *Wem gehört die Stadt? Armut und Obdachlosigkeit in den Metropolen.* Basel, Switzerland: Lenos.

Body-Gendrot, S. (in press). Marginalization and political responses in the French context. In P. Hamel, H. Lustiger-Thaler, & M. Mayer (Eds.), *Urban fields/global spaces: The phenomenon of urban movements in a global environment.* Thousand Oaks, CA: Sage.

Boll, J., Mauthe, A., & Osoriao, M. (1991). Stattbau GmbH Hamburg. In R. Froessler & K. Selle (Eds.), *Auf dem Weg zur sozial und ökologisch orientierten Erneuerung?* (pp. 225-256). Dortmund, Germany: Dortmunder Vertrieb für Bau- und Planungsliteratur/Wohnbund.

Boyte, H., Sirianni, C., Barber, B., & Delli Priscoli, J. (1994). *Reinventing citizenship project: A proposal.* Proposal, Humphrey Institute of Public Affairs, in collaboration with the White House Domestic Policy Council.

Brown, P., & Crompton, R. (Eds.). (1994). *Economic restructuring and social exclusion.* London: UCL Press.

Bullard, R. (Ed.). (1993). *Confronting environmental racism: Voices from the grassroots.* Boston: South End.

Castells, M. (1983). *The city and the grassroots.* Berkeley: University of California Press.

Castells, M. (1994). European cities, the informational society, and the global economy. *New Left Review, 204,* 18-32.

Clarke, S. (1994, August). *Rethinking citizenship: The political implications of the feminization of the city.* Paper presented at the ISA World Congress, Bielefeld, Germany.

Clarke, S. E., & Mayer, M. (1986). Responding to grassroots discontent: Germany and the United States. *International Journal of Urban and Regional Research, 10*(3), 401-417.

Cortés, E., Jr. (1996). Reweaving the social fabric. *Boston Review.* (Reprinted in *FOCO* [Forum for Community Organizing], March 1997, pp. 16-20)

Cress, D. M., & Snow, D. A. (1996). Mobilization at the margins: Resources, benefactors, and the viability of homeless SMOs. *American Sociological Review, 61,* 1089-1109.

Dangschat, J. (1995). "Stadt" als Ort und Ursache von Armut und sozialer Ausgrenzung. *Aus Politik und Zeitgeschichte, 31/32,* 50-62.

Davis, M. (1990). *City of quartz: Excavating the future of Los Angeles.* London: Verso.

Demirovic, A. (1998). NGOs and social movements: A study in constrasts. *Capitalism, Nature, Socialism: A Journal of Socialist Ecology, 9*(3), 83-92.

Di Chiro, G. (1992). Defining environmental justice: Women's voices and grassroots politics. *Socialist Review, 22*(4), 93-130.

Dreier, P. (1993). *Community empowerment strategies: The experience of community-based problem-solving in America's urban neighborhoods—Recommendations for federal policy.* Roundtable on Urban Policy, U.S. Department of Housing and Urban Development, Washington, DC.

Egan, T. (1993, December 12). In three progressive cities, stern homeless policies. *The New York Times,* p. 26.

Evers, A., & Olk, T. (Eds.). (1996). *Wohlfahrtspluralismus: Vom Wohlfahrtsstaat zur Wohlfahrtsgesellschaft.* Opladen, Germany: Westdeutscher Verlag.

Fehse, W. (1995). *Selbsthilfe-Förderung: "Mode" einer Zeit? Eine Prozeß- und Strukturanalyse von Programmen zur Unterstützung von Selbsthilfeaktivititäten.* Frankfurt, Germany: Peter Lang.

Ferguson, S. (1991, June 18). The park is gone. *The Village Voice,* p. 25.

Fiedler, J., & Schrödter, R. (1983). Der Zweite Arbeitsmarkt in Hamburg: Nützliche Beschäftigung statt Hinnahme und Finanzierung von Langfristarbeitslosigkeit. In M. Bolle & P. Grottian (Eds.), *Arbeit schaffen—jetzt* (pp. 165-186). Hamburg, Germany: Rowohlt.

Fink, U. (Ed.). (1983). *Keine Angst vor Alternativen: Ein Minister wagt sich in die Szene.* Freiburg, Germany: Herder.

Fisher, R., & Kling, J. (Eds.). (1993). *Mobilizing the community: Local politics in the era of the global city.* Newbury Park, CA: Sage.

Fishman, N., & Phillips, M. (1993). *A review of comprehensive, collaborative, persistent poverty initiatives.* Evanston, IL: Northwestern University, Center for Urban Affairs and Policy Research.

Froessler, R., Lang, M., Selle, K., & Staubach, R. (Eds.). (1994). *Lokale Partnerschaften: Die Erneuerung benachteiligter Quartiere in europäischen Städten.* Basel, Switzerland: Birkhäuser.

Grottian, P., Krotz, F., Lütke, G., & Pfarr, H. (Eds.). (1988). *Die Wohlfahrtswende: Der Zauber konservativer Sozialpolitik.* Munich: Verlag C. H. Beck.

Hall, T., & Hubbard, P. (Eds.). (1998). *The entrepreneurial city.* Chichester, UK: Wiley.

Harris, L. (1995, July 24). Banana Kelly's toughest fight. *The New Yorker,* pp. 32-40.

Harvey, D. (1998). Globalization and the body. In International Network of Urban Research and Action (Ed.), *Possible urban worlds: Urban strategies at the end of the 20th century* (pp. 26-39). Basel, Switzerland: Birkhäuser.

Häußermann, H., & Siebel, W. (Eds.). (1993). Festivalisierung der Stadtpolitik [special issue]. *Leviathan, 13.*

Hirsch, J., & Roth, R. (1986). *Das neue Gesicht des Kapitalismus.* Hamburg, Germany: Argument Verlag.

Holusha, J. (1994, May 6). Pioneering Bronx plant to recycle city's paper. *The New York Times,* pp. D1, D5.

Hopkins, E. (1995, Spring). At the cutting edge: A portrait of innovative grassroots organizations in Los Angeles. *Critical Planning, 2,* 39-60.

Huster, E.-U. (1997). Zentralisierung der Politik und Globalisierung der ökonomie: Veränderungen der Rahmenbedingungen für die soziale Stadt. In W. Hanesch (Ed.), *Berblet die Soziale Stadt? Konzeption, Krise, und Perspektiven kommunaler Sozialestaatlichkeit* (pp. 57-75). Oplanden, Germany: Leske & Budrich.

Jacobs, B. D. (1992). *Fractured cities: Capitalism, community, and empowerment in Britain and America.* London: Routledge.

Kingsley, G. T., McNeely, J. B., & Gibson, J. O. (1997). *Community building coming of age.* Washington, DC: Urban Institute.

Kramer, J. (1988, November 28). Letter from Europe. *The New Yorker,* pp. 67-100.

Knorr-Siedow, T., & Willmer, W. (1994). *Sozialverträglicher Umgang mit unkonventionellen, mobilen Wohnformen am Beispiel des Wohnens in Wohnwagendörfern oder Wagenburgen.* Berlin: IRS.

Krämer-Badoni, T. (1990). Die Dethematisierung des Sozialen: Ansätze zur Analyse städtischer sozialer Bewegungen. *Forschungsjournal Neue Soziale Bewegungen, 4,* 20-27.

Krämer-Badoni, T., & Söffler, D. (1994). *Die Rolle der städtischen Bürgerinitiativen in Westdeutschland und Ostdeutschland bei der Ausprägung lokaler Demokratie* (pp. 67ff). Working Papers of the ZWE Arbeit und Region, No. 13, Universität Bremen.

Krätke, S. (1991). *Strukturwandel der Städte: Städtesystem und Grundstücksmarkt in der "post-Fordistischen" Ära.* Frankfurt, Germany: Campus.

Krätke, S. (1995). *Stadt–Raum–Ökonomie.* Basel, Switzerland: Birkhäuser.

Lang, M. (1994). Neue Handlungsansätze zur Erneuerung benachteiliger Stadtquartiere in Deutschland. In R. Froessler, M. Lang, K. Selle, & R. Staubach (Eds.), *Lokale Partnerschaften: Die Erneuerung benachteiligter Quartiere in europäischen Städten* (pp. 161-175). Basel, Switzerland: Birkhäuser.

Lang, S. (1995). Civil society as gendered space: Institutionalization and institution building within the German women's movement. In J. W. Scott & C. Caplan (Eds.), *Translations, environments, transitions: The meanings of feminism in changing political contexts* (pp. 101-120). New York: Routledge.

Logan, J. R., & Molotch, H. L. (1987). *Urban fortunes: The political economy of place.* Berkeley: University of California Press.

Lustiger-Thaler, H., & Maheu, L. (1995). Social movements and the challenge of urban politics. In L. Maheu (Ed.), *Social movements and social classes* (pp. 151-168). London: Sage.

Matzdorf, R. (1989). Beschäftigungsinitiativen und ihre Förderung in Nordrhein-Westfalen. In C. Mühlfeld, H. Oppl, H. Weber-Falkensammer, & W. R. Wendt (Eds.), *Brennpunkte sozialer Arbeit* (pp. 41-50). Frankfurt, Germany: Diesterweg.

Mayer, M. (1987a). Restructuring and popular opposition in West German cities. In M. P. Smith & J. R. Feagin (Eds.), *The capitalist city: Global restructuring and community politics* (pp. 343-363). Oxford, UK: Basil Blackwell.

Mayer, M. (1987b). Städtische bewegungen in den USA: "Gegenmacht" und Inkorporierung. *Prokla, 17*(3), 73-89.

Mayer, M. (1993). The career of urban social movements in West Germany. In R. Fisher & J. Kling (Eds.), *Mobilizing the community* (pp. 149-170). Newbury Park, CA: Sage.

Mayer, M., & Katz, S. (1985). Gimme shelter: Self-help housing struggles within and against the state in New York City and West Berlin. *International Journal of Urban and Regional Research, 9*(1), 14-46.

McCarthy, J. D., & Castelli, J. (1994). *Working for justice: The campaign for human development and poor empowerment groups.* Washington, DC: Catholic University of America, Life Cycle Institute.

Mette, N., & Steinkamp, H. (Eds.). (1997). *Anstiftung zur Solidarität: Praktische Beispiele der Sozialpastoral.* Mainz, Germany: Matthias-Grünewald Verlag.

Ministry of Social Affairs. (n.d.). *Appeal to civic engagement in Baden-Württemberg.* Available on Internet: www.aktiv.de/buerger.

Mollenkopf, J., & Castells, M. (1991). *Dual city: Restructuring New York.* New York: Russell Sage.

Natz, S. (1993, October 25). Furcht vor weiteren Räumungen: Bunter und friedlicher Protestzug von rund 700 Wagendorfbewohnern und Hausbesetzern. *Berliner Zeitung.*

Péchu, C. (1996). Quand les "exclus" passent à l'action: Ma mobilisation des mal-logés. *Politix, 34,* 114-133.

Pickvance, C. G. (1995). Where have urban movements gone? In C. Hadjimichalis & D. Sadler (Eds.), *Europe at the margins: New mosaics of inequality* (pp. 197-217). London: Wiley.

Piven, F. F., & Cloward, R. A. (1977). *Poor people's movements: Why they succeed, how they fail.* New York: Pantheon.

Rich, M. (1995, April). *Empower the people: An assessment of community-based, collaborative, persistent poverty initiatives.* Paper presented at the annual meeting of the Midwest Political Science Association, Chicago.

Rivera, Y., & Hall, J. (1996, July 4). *Global cities and grassroots democracy: The case of Banana Kelly.* Lecture given at the Amerika-Haus Berlin.

Roth, R. (1994). *Demokratie von unten: Neue soziale Bewegungen auf dem Wege zur politischen Institution.* Bonn, Germany: Bund Verlag.

Rucht, D., Blattert, B., & Rink, D. (1997). *Soziale Bewegungen auf dem Weg zur Institutionalisierung: Zum Strukturwandel "alternativer" Gruppen in beidenteilen Deutschlands.* Frankfurt, Germany: Campus.

Ruddick, S. (1994). Subliminal Los Angeles: The case of Rodney King and the sociospatial reconstruction of the dangerous classes. *Gulliver,* No. 35 ("Die neue Metropole: Los Angeles-London"), 44-62. (Hamburg, Germany: Argument Verlag)

Russell, D. (1990, Winter). The rise of the grassroots toxics movement. *Amicus Journal, 12,* 17-21.

Sambale, J. (1994). *Fluchtpunkt Los Angeles: Zur Regulation gesellschaftlicher Beziehungen über lokale Umweltpolitiken in einer internationalen urbanen Region.* Unpublished thesis, Free University of Berlin.

Sassen, S. (1997, October). *Whose city is it? Globalization and the formation of new claims.* Paper presented at the meeting of the German Political Science Association, Bamberg, Germany.

Scholz, C. (1997). Überall ist Mega Mall: Stadtentwicklung, Strukturwandel, und der Wettlauf der Erlebniswelten. *AKP-Fachzeitschrift für Alternative Kommunalpolitik, 18*(5), 32-35.

Schubert, D. (1990). Gretchenfrage Hafenstraße: Wohngruppenprojekte in Hamburg. *Forschungsjournal Neue Soziale Bewegungen, 3*(4), 35-43.

Selbsthilfe ist gefragt: Bundes-Modellversuch förderte Bildung neuer Initiativen. (1992, October 23). *Frankfurter Rundschau.*

Selle, K. (1991). *Mit den Bewohnern die Stadt erneuern: Der Beitrag intermediärer Organisationen zur Entwicklung städtischer Quartiere—Beobachtungen aus sechs Ländern.* Dormund, Germany: Vertrieb für Bau- und Planungsliteratur/Wohnbund.

Servon, L. J. (1997). Microenterprise programs in U.S. inner cities: Economic development or social welfare? *Economic Development Quarterly, 11*(2), 161-180.

Shragge, E. (Ed.). (1997). *Community economic development: In Search of empowerment.* Montreal: Black Rose Books.

Shragge, E., & Fontan, M. (1996). *Workfare and community economic development in Montreal: Community and work in the late twentieth century.* Unpublished manuscript, School of Social Work, McGill University (Shragge), and Université du Quebec à Montréal (Fontan).

Smith, N. (1996). *The new urban frontier: Gentrification and the revanchist city.* London: Routledge.

Stiftung Mitarbeit. (Ed.). (1995). *Engagement gegen soziale Not.* Bonn, Germany: Author.

Tageszeitung. (1998, February 3).

Vogelfrei. (1994, September).

von Hoffmann, A. (1997, January). Good news! From Boston to San Francisco, the community-based housing movement is transforming bad neighborhoods. *Atlantic Monthly,* pp. 31-35.

Wagner, D. (1993). *Checkerboard Square: Culture and resistance in a homeless community.* Boulder, CO: Westview.

Wright, T. (1997). *Out of place: Homeless mobilizations, subcities, and contested landscapes.* Albany: State University of New York Press.

Zukunft Bauen et al. (1994). *Zukunft Bauen 1993-1994.* Berlin: Geisel Druck.

Part IV

Prescriptive Visions

11
(Promised) Scenes of Urbanity

CHRISTIAN RUBY

Faced with the tragedies of urban violence and growing misery, more and more people find it unacceptable to limit urban politics to the simple management of public problems, to functional or territorial management. Submitting this to a philosophical frame of mind, we need to question our concepts of "the urban" to understand what the idea of the city has to offer. Ideally, it will provide conceptual resources and practical guides revealing new ideas and forging a new moment in urban history.

Before entering such a discussion, we must recognize that public attention has mainly focused on two models of the organization of the city. Yet, these are barely analyzed, and when they are, they promote only fear or admiration. If we do not study such models, then we will be unable to understand how they emerged from a long crisis of reorganization that rejected the Fordist way of thinking. Consequently, obstacles to the comprehension of today's urbanity (by which we mean the shape and social linkage of morals, daily life in the city, and social contacts within the city's boundaries) will be perpetuated.

The choice is between the postcrisis city, which leads to groups of people living in distinctive parts of town locked in a so-called autarky that has emerged from a dual society, and (if we refuse this model) a city caught by its past, managed as if it were an ancient metropolis focused on the celebration of common abstract values and state experiences whose monuments always have reflected its requirements. These models hardly advance the development of an urban civilization by which people experience a life jointly organized.

The (promised) scenes of urbanity that we would like to evoke—moving among images, realities, and hopes—are mostly driven by the desire to excavate proposals that enable citizens to elude an urban politics led by experts, a politics that deprives citizens of their abilities to decide and act. Only in this way can citizens regain control over urbanity.

False Accusations but True Failures

In the first scene are images of urban public space. We start with this scene because it defines the conditions that make possible our sighting of the city—how we are touched by its good and bad sides. Typically in this scene, few people distinguish the actual obstacles that confront them, the main one being that the "city" is now less a reality than a reactive image. Admittedly, it is not easy to grieve the city, but if we want to contribute to an open space of public discussion such as that proposed by Jürgen Habermas concerning urbanity (Habermas, 1997), then we must confront this problem immediately.

Urbanity, as we know, does not express itself only through its spatial and material aspects. It obeys logics of representation—images it allows of itself, images it shows to itself, images we build of it, images it refuses. Although it is not necessary to be precise with the most common images—they change unceasingly—we cannot neglect the fact that they have a hold on us. Thus, we consider the city as a place for sociability, as a system of relations between different spaces, as a simple network of uncontrollable fluxes, as an imposed way of life, and/or as a moment of liberation of humankind, all of which influence our appreciation of forthcoming transformations. If today the stilted myths of a postmodern babel or of a hypermodern global city also appear, then it only confirms our first impression. These images, however, are not always best for explaining the new stakes of our times, that is, for enabling citizens to speak freely and embrace responsibility for their actions and lives or for calmly anticipating the city of the future.

In regard to these representations, the dominant feeling is of an incapacity to reject two extreme images of urbanity: the urban plans of the metropolitan era that focused on technical solutions to population growth and the sensations experienced by the urban stroller. Because they are diametrically opposed, these images are identical.

On the one hand, we can evoke an immense collection of archives that extends from the painter Edward Hopper to the novelist Sinclair Lewis, from the film director Fritz Lang to the sculptor Arman, or even

Jean-Paul Sartre's description of New York City. These views combine suspicion and hope and are linked to cities of great population size and their social preoccupations.

On the other hand, we experience annoyances and pleasures that appear to us through architectural nostalgia, best shown in movies such as Terry Gilliam's *Brazil* and Emir Kusturica's *Underground* or in various musical shows that present themselves as "Portes ouvertes sur ville" (Luc Ferrari). At times, this second set of views is expressed through the sentimental theme of symbolic ruins, a nostalgia for a lost unity and for past forms that once, ostensibly, symbolized absolute values.

Most of the talk about the interconnection of art and the city clings to these symmetric couplings of two opposite (but identical) images of urbanity. The question is whether we should continue to use, through its representatives and heroes, public art in public spaces to display the unity of the political body. This question was answered one way during the 19th century. Public monuments had to offer the image of a collective future by being social landmarks to which people could refer symbolically and where they could meet and hold ceremonies—places such as Times Square and the entryway of the Guggenheim Museum in New York City. Yet, as surprising as it might be, people continue to think in the same way; some assert that we must rearticulate the ancient significance of monuments, others ask artists to find new symbols for contemporary times, and still others believe that a postmodern crumbling is erasing the ancient in favor of new forms of leisure, even though these forms would occupy the same sites as monuments to ancient regimes.

Nobody sees that we must cast aside the opposition between maintenance and nostalgic compensation. We are of a time that requires the invention of a new concept of the urban if we are to understand what is happening before our eyes and construct new proposals. The most common image of the city is its cries—grafitti, hip-hop, urban dancing—built on the status of victims, swept up by an uncontrolled history, and spread by the absence of a new language.

■ Sociospatial Configurations

For this second scene, we analyze how the dual society and its urbanity are represented. Each urban form, of course, is the result of historical and political trajectories that make it impossible to think about the urban without the social conditions that support it. For now, we live in societies in which social mobilities have stalled. Few people can climb the social ladder, as witnessed by the spread of ghettos. Furthermore, the retreat of social mobility stifles spatial stimulations. Urban forms turn into static figures where everything is motionless. This stasis is reinforced by security systems that deploy electronic frontiers and watchdogs.

Nevertheless, ancient cities cannot be praised simply because they were different. They had their own ways of assuming responsibilities in the social field, especially through the modification of urban space—reorganization, revaluation, reinstatement, and renovation. Their topographies experienced constant metamorphosis with the necessary synthesis. "Each of these large divisions of Paris [was] a city, but a city too special to be complete, a city that could not manage without the two others" (Hugo, 1969, p. 129). Furthermore, modern urbanism, more than one of a city of entrapment ("Mahagonny") from which people cannot escape physically or mentally, did not pretend to do anything else but to move on with, or even contribute to, social promotions (*La Belle equipe*). London, New York, Berlin, and Paris never stopped offering spatial equivalences to the welfare state, whose mission was to redistribute social opportunities as well as urban perspectives. Spatial movements accompanied the resultant social integration and the various forms of democratic assimilation. The more a group moved within a social context, the more it became spatially mobile and the more its spatial concentration was diluted.

Still, sociospatial mobilities have stalled to such a point that urban dynamics are interrupted and important separations in urban space, parts of town that heretofore had been closed, are extended through expropriations and demolitions. After the technopoles (i.e., industrial districts) were pushed out and residential areas and living quarters were separated, fewer and fewer regulations were promoted and social confrontations decreased. The future of too many people demands the breaking of the solidarity that they share with others. In a way, only the *laisses pour comptes* benefit from mobility, but on trajectories that even road movies avoid because they cannot identify this type of mobility. We are not describing here anything but the dual city, the physical manifestation of a society that locks social groups into enclosed territories and obliges them to live at two different speeds. They are defined as static statuses rather than by their connections to one another; they are included by addition, not relation.

This configuration does not show up in the city's history. That history presumes social dynamics that can be detected. What most reveals this configuration and its history are the various forms of resistance that pervade them. Do these generate the possibility of reopening the city? The *Wagenburgler* in Berlin, the *squatters* in New York City, and the *exclus* in Paris must no longer be treated as objects that exist solely to satisfy the charitable impulses of the wealthy. With their possibilities, they resist the political and social connections being built, even though this resistance does not find an equivalence in a public space strangely deaf to increasing animosities. Municipalities hope to eradicate these havens for rioters, but we must ask why those who are "set aside" should not also play a part in the city.

■ Urban Solidarities

In the third scene, with the dual city, the fate of an active political life is doomed and sealed. Urban solidarity has been replaced by a general feeling of being at odds and a sense of discomfort. We are imprisoned in selfishness. These worries are legitimately felt by everybody. Everywhere, scatterings are multiplying—wars separating populations, unemployment shattering relations, destructions emptying memories, pollutions killing, corruption breaking social links. Silences and common solitudes exacerbate these fragmentations. Within these scatterings, the cities have various roles. Some cities are being destroyed by people who hate their modern functions, other cities are trying to maintain pace with evolving urban lifestyles, and still other cities are attempting to adapt to a new urban environment.

In this context, a few cities act as places for refugees (refuge cities) and hope to maintain some level of solidarity. Other cities have exiled writers and artists—Berlin, Strasbourg, Carn, Almeria, Valladolid, Göteborg, São Paulo, Gorea, Nagasaki, Durban. In the same way as people attempt to relegitimate their surroundings, cities attempt to reintegrate themselves to maintain a hold on their futures. They deploy urban designers with bold visions, even though our times are inhospitable to their proposals.

Pleasant and useful to everybody who dares to listen, solidarity presents itself as unequivocally distinct from these separations. Nevertheless, it still is too vague. Too many people use it to achieve contradictory goals or mistake it for mere compassion. People use the word *solidarity* abusively. Solidarity cannot be what attraction is to planets—a natural fact. Solidarity is not given to people; it has to be commonly forged within a specific environment and according to current and past events. Solidarity will grow to universal size only if it includes a political savoir faire, a way of seeing that lies in the urban world and brings together people in action. Frontier cities such as Skopje, Salonika, Prague, and Bratislava demonstrate that the road from an urban coexistence (different ethnic groups, different cultures) to solidarity is a difficult one.

From a philosophical point of view, we must offer citizens new ideas and elaborate new political projects or at least help to structure multiple longings for a society, as well as a world, with a different urban shape. So, what generates solidarity? When does it become effective? There is no ambiguity in this. If it is possible to speak of a culture of urban solidarity, then it is because solidarity comes from a political culture that pursues the common good rather than the dispersed, allows public debates rather than controlling information, and tolerates wanderings and cities without banishments. Because separation and resentment have become dominant political and urban qualities, we must construct tools that enable us to anticipate and propose, not react. Only then can we imagine a future that is socially and politically true.

In this frame of mind, the victim is the major witness of a political thinking set in a world that still is postmodern. This notion reminds us that without criticism, the urban and social world will produce victims—citizens acting in closed worlds that violently reject the worlds of others, resisting invasion or fighting among themselves.

■ Archipelagos (of Invention)

Our fourth scene is self-defined—the development of a political culture of solidarity that embraces new practices and enables urban residents to rediscover their citizenship. Times of crisis, like the ones we are experiencing, not only reveal systems but also activate initiatives that weaken public engagement. They remind us that citizenship is not only a right but also a collective action.

In the most devastated urban zones, where the explusion of public responsibilities grows in parallel with violence and delinquency, collective work (e.g., health care, social work, scholarly interventions, complementary [job] training) has led people to push for local democracy. Their inhabitants, professionals, members of government, and local associations have discovered new means of dialogue and mutual aid.

Through such examples, we can link our ideas to the notion of an archipelago. This notion designates the means of action for today. It offers propositions for existence that philosophy claims and presents to all. It delimits a range of solidaristic practices—by gender, by age, by city quarter, by workplace—that compensate for a dead state with actions that promote new and dynamic social forms.

The archipelago promotes active links, whether existing or potential, among citizens who wish to embrace their hopes and who are willing to act together to accomplish a common history. It creates new limits for experiments with which citizens can regain sovereignty over their existence, of which they are deprived today. As such, the archipelago refers to collective actions, recognizing that action cannot be decreed or imposed. It must be felt and conquered. Within its definition, all citizens have to face both themselves and their relations toward others. Only this engagement gives full meaning to the notion of "acting in the archipelago."

Of course, our list of references is incomplete. We presented a few examples to delineate the common emptiness of the subject. There is too much talk about network technologies (e.g., cable multimedia, wireless communications) as solutions to urban problems or, alternatively, to fears of destruction. Here again, if we wish to improve the situation, we must learn to avoid debates that are overly simplistic and must basically oppose technocracy to technophobia. That is why the use of the archipelago notion is so useful; it prevents us from considering urban residents as motionless characters set before dominant communication forms. Instead, it places them in the midst of citizen interactions and habits.

The archipelago is not a lifeless or static concept, a picture of what is happening, but rather a draft of what could be, a new way of thinking and living in the city. The notion of archipelago inserts a logic into the whole of social and urban dynamics and encourages common attitudes meant to resist the fragmentation of the global order.

Importantly, if we want citizens to be responsible for their futures, then room must be left for controversy. With the archipelago notion, we hope to fight the forces that impose reunification and result in an abstract urbanism and the forces that create the desperate dispersions of a postmodern urbanism. We prefer deep reflection about the different linkages among various actions and about the affinities among social beings with common habits. Today, nobody has to declare himself or herself a part of these forces—the almost transcendent traditional principles of unity or of the harmony of the diverse or the synthesis of the multiple, the system, the encyclopedia or organism, the underlying correspondence of analogy. Although these principles are not as coercive as the postmoderns claim, they are presented with the authority of abstract identity or the authority of the tyranny of the end over the means or even of the end of history—the forever happy city. We do not have to succumb to dispersion or to the close-in tendencies that finally conclude in their opposite.

Yet, the notion of the archipelago must make explicit the authority of an infinite movement that coordinates common actions that constantly are rebuilt, a "doing together" actualized, not a "being together." In this sense, the notion has a front-row seat in the new building of a conscience of history but not in a policy of consensus or in an aesthetic of dispersion. Because we do not want to attribute a final meaning to history, the notion of archipelago resists a laissez-faire attitude.

Between the policy of closed and added quarters—Lego-like constructions that can be added and subtracted, moved about, and replaced and that never are truly connected—and the blind maintenance of all in an inherited form lies an open space for thought, a space for thinking about new proposals for the city of the future. Instead of a patchwork of societies, we must look for a city based on relationships, a city structured as an archipelago and anchored in the expansion of solidarity.

REFERENCES

Habermas, J. (1997). *Droit et morale.* Paris: Seuil.

Hugo, V. (1969). Paris a vol d'oiseau. In *Notre Dame de Paris* (Vol. 1, Book 3, Chapter 2). Paris: Éditions de L'Érable.

12

Can We Make the Cities We Want?

SUSAN S. FAINSTEIN

The profession of city planning was born of a vision of the good city. Originating in 19th-century radicalism and utopianism, its program responded to the evils of the industrial city (Fishman, 1977; Hall, 1988). City planners aimed at creating a city in which the insalubrious environment and social structure of industrial capitalism would be defeated by a reordering of physical and social arrangements, even while its bounty continued. The hope was that all citizens could attain the benefits of beauty, community, and democracy not through revolutionary means but rather through the imposition of reason.

Today, planning practitioners conceive their mission more modestly. The concept of visionary leaders imposing their views on the urban populace is in disrepute, whereas the notion of an identifiable model of a good city meets skepticism. The aim of creating the good city could not maintain itself against the intellectual and practical onslaughts of the succeeding century. Both premises—of a unified vision of the good city and of a disinterested elite that could formulate and implement it—fell before attacks from across the ideological spectrum (Fainstein & Fainstein, 1996; Fischer, 1990; Lucy, 1988). In the crudest form, people on the left argued that elites under capitalism always would represent the interests of the upper class (Harvey, 1985), whereas thinkers on the right contended that any effort by the state to direct urban outcomes would constitute an infringement on liberty (Hayek, 1944) and a deterrent to efficiency (Klosterman, 1985). To put it another way, leftists disputed the possibility of finding a common good for a class-divided society, whereas rightists feared that state intrusion would interfere with individual choice and market allocations. Meanwhile, more centrist thinkers focused on the inadequacies of planning processes and methodology, arguing that comprehensive planning was inherently undemocratic and unattainable (Altshuler, 1965; Lindblom, 1959). Thus, this mainstream view implied that identification of a collective vision of the good city was not possible. At the same time, planning practice, as it became embodied in exclusionary zoning, urban renewal, highway construction, and public-private partnerships, apparently confirmed the jeremiads of its critics (Anderson, 1964; Gans, 1968; Squires, 1989; Toll, 1969).

Nevertheless, although their critique of traditional planning undermined the foundations of its normative framework, critical urban theorists continued to harbor an ideal of a revitalized, cosmopolitan, just, and democratic city. Most did not explicitly justify their value criteria; instead, in the words of the U.S. Declaration of Independence, they simply deemed certain truths to be self-evident.[1] Recently, however, there has been a revival of discussion of the appropriate values that ought to govern urban life.[2] These contemporary efforts, with varying success, have had to take account of the postmodernist/poststructuralist assault on the existence of a unitary ethic and its emphasis on the situatedness of the speaker. Thus, recent work has attempted to recognize the divided, perceptual nature of social concepts of the good while laying out a broad common value structure that embraces difference itself.

This chapter extends this contemporary discussion of what type of city we want, looking in particular at the extent to which it is possible to combine the values of equality, diversity, participation, and sustainability in a cosmopolitan urban milieu. The question is, in part, an exploration of the potential for urban planning, where the term refers to the conscious formulation of goals and means for metropolitan development, regardless of whether these determinations are conducted by people officially designated as planners or not. Underlying its argument is the premise that a city should be purposefully shaped rather than the unmediated outcome of the market and of interactions within civil society, in other words; that planning is a necessary condition for attaining urban values.[3] The next section examines the normative framework of contemporary progressive urban theory and presents the arguments within some recent writings concerned with urban values. After that, the cases of Amsterdam and New York are briefly presented to reveal the forces that enable or block the realization of these values. Finally, I argue that although one can—and should—erect a pantheon of values, the strategies for achieving them and the priorities to be accorded to each vary according to historical context.

Values and Urban Theory

Urban ecology, which initially was the dominant approach to urban studies, pretended to scientific neutrality but, in its naturalizing of social processes, supported a politically conservative outlook.[4] Later, amid the social movements of the 1960s, a new generation of scholars appeared who chose to analyze the impact of urban development on class, race, ethnicity, democratic decision making, and environmental sustainability. Rather than accepting the city as the outgrowth of natural processes, they regarded it as a social construction resulting from the exercise of power. Similarly, a new generation of urban practitioners eschewed the neutrality of their predecessors and identified with the cause of the underdog. They acted as guerrillas in the bureaucracy and advocates for community groups rather than as dispassionate experts (Davidoff, 1965; Needleman & Needleman, 1974). Nevertheless, although the analyses and activities of these new groups of scholars and practitioners were overtly value laden, their work was critical and oppositional rather than transformational; the value criteria they used typically were assumed rather than explicated. Generally, however, we can say that underlying the tradition of critical urban scholarship and practice that established itself during the 1970s and continues to the present is a commitment to social justice variously defined.[5] Essentially, the achievement of social justice requires that existing groups have equal access to material well-being, symbolic recognition, and decision-making power and that future generations inherit an environment that has not seriously deteriorated. To put this another way, social justice is based on material equality, social diversity, democracy, and environmental sustainability (Fainstein, 1997b).

Theorists tend to gloss over the problems of realizing these values simultaneously. For example, the value of democracy is particularly problematic when democratic participation results in exclusion and the preservation of privilege. Wealthy suburbs use zoning to keep out potential purchasers of inexpensive housing; low-income, white, inner-city neighborhoods exclude different racial groups, often with implicit threats of violence. The tension among democracy, diversity, and equity is obvious in these situations. In a different example, leaders of racial minorities focus on the recognition of group rights and symbolic recognition. At the same time, they might ignore economic processes and tolerate income inequality so long as members of their group have an equal shot at whatever opportunities exist. The result can be perpetuation or even worsening of inequality despite greater diversity.

Efforts at reconciling these beneficent yet discordant principles become further complicated when other widely desired aims also are accommodated. The conservative values of order, efficiency, and economic growth clash with those of equality and diversity, at least in the short run. Although thinkers on the left might dismiss the former set of values as simply supportive of the status quo and legitimated through propaganda, transgressing it remains anti-democratic if such transgression means suppressing popular majorities. Even environmentalism, which appears to be the strongest oppositional grassroots popular movement affecting cities and regions, fits uneasily at best into the rest of the left program, as the difficulty of sustaining "red-green" political coalitions in European cities attests. The tension here derives precisely from the priority given to employment and economic growth among those for whom economic equality is the highest value, as opposed to the "small is beautiful," preservationist, quality-of-life emphasis of the environmentalists.

A recent body of work has begun to address the issue of the good city directly, although its thrust has not been to address the tensions among value commitments. Rather, theorists have sought to deal with the difficulties of offering prescriptions for urban betterment without falling into the absolutism that characterized their much-criticized predecessors. In so doing, they have responded to the postmodernist attack on Enlightenment thought. For some thinkers, the effort at escaping a totalizing discourse and rule by an elite has resulted in a definition of the good city as a process rather than as a particular outcome. For others, most notably David Harvey, ultimate condition matters more than how it is achieved (see also Lauria, 1996). I summarize some of these efforts to illuminate the problems of attempting to establish norms while remaining conscious of the partiality of one's own viewpoint. My choice of works is necessarily arbitrary; they represent what I regard as the most provocative of recent writings specifically concerned with the urban realm.

Critical Theorists and Urban Values

Critical urban theorists share a repudiation of certain of the methods and values of conventional social scientists and a common commitment to social justice. They criticize positivist methodologies as distorting and superficial, question typical measures of benefits such as aggregate growth and consumer choice, and support the claims of the socially marginalized and oppressed. They largely acknowledge that social science traditionally has universalized the viewpoint of the white Western male. At the same time, they differ among themselves on a number of issues, most notably the sustainability of Enlightenment-based concepts of reason, the relative importance of class vis-à-vis other identity groupings, the possibility of locating normative systems that transcend cultural particularities, and the usefulness of utopianism. Although some adopt the postmodernist viewpoint of irreducible subjectivity, others accept its attack on previous efforts at universalism yet still are committed to discovering overarching values.

Harvey confronts head-on the postmodernist challenge to political economy in several of his works. He also deals with the issue of environmentalism, incorporating it into his general argument concerning distributional justice. In *The Condition of Postmodernity,* Harvey (1989a) brilliantly develops the postmodernist argument concerning the inadequacies of modernist discourse, using many of the techniques of cultural studies (e.g., textual analysis, references to film and fiction) to substantiate his argument. He acknowledges the importance of history, gender, race, and ethnicity in directing action and creating meanings. He convincingly shows that postmodernism is a further development of modernism, which always recognized the fragmentary, ephemeral quality of life (Berman, 1988). But in continuing to cling tenaciously to his Marxist underpinnings, he never gives up attributing ultimate explanatory power to economic structure and ultimate value priority to economic equality. Similarly, in his most recent book, *Justice, Nature, and the Geography of Difference,* Harvey (1996) critiques the management-of-nature approach that has characterized nearly all post-Enlightenment thought including Marxism. He then slips from a problematizing of the tension between human comfort and respect for the environment to a defense of environmental justice that shifts the focus from the relations of humans to nature to the relations among human groups.

Essentially, for Harvey, economic equality is so fundamental to all forms of justice that, although he raises profoundly important questions that seem to lead to other criteria of value, he always returns to the same point of judgment: Capitalism is the source of social injustice. In a summary essay on social justice, he states,

> Only through critical reengagement with political-economy, with our situatedness in relation to capital accumulation, can we hope to reestablish a conception of social justice as something to be fought for as a key value within an ethics of political solidarity built across different places. (Harvey, 1996, p. 360)

When dealing with conflicts among identity groups or, in fact, with any issue of social power, Harvey invariably regards power as an outcome of ownership of the means of production. Thus, he minimizes the significance of noneconomic forms of subordination and refuses to accept that economic dominance can be a consequence of power rather than the cause of it.

The great strength of Harvey's analysis is that it never loses sight of the implications of economic inequality and the material obstacles to overcoming dominance hierarchies. Like all the thinkers described here, Harvey finds injustice in the social construction of subordination. Unlike most of them, however, he remains a structuralist and, therefore, considers hierarchy as deeply embedded within economic relations and not susceptible to uprooting through symbolic activity. He perhaps underestimates the importance of perception in propping up economic structures and, therefore, the role of discourse in maintaining the status quo. Still, he does not indulge in either the utopianism or the naive exhortation that seems to characterize much of contemporary theory. Although social discourse can be "read" to discover underlying inequalities, reveal injustice, and demystify hierarchy, rewriting the discourse does not in itself reconstitute the structure. Thus, in the typical case, deconstruction does not even result in a reconstituted text. As a recent study of planning in a Danish city attests, critical exposure of convenient rationalizations by those in power will not succeed, on its own, in undermining prevailing myths:

> Power determines what counts as knowledge, what kind of interpretation attains authority as the dominant interpretation. Power procures the knowledge which supports its purposes while it ignores or suppresses that knowledge which does not serve it. (Flyvbjerg, 1998, p. 226)

Unlike Harvey, John Friedmann embraces utopianism but characterizes his particular form of it as realistic (Friedmann, 1987, p. 343).[6] Although he mentions equality as a necessary precondition to human fulfillment and condemns the dehumanizing effects of capitalism, his emphasis is on activity, not condition. He calls for transformative planning within the context of a movement toward the recovery of political community. The latter is defined in terms of a set of relational goals—widespread citizen involvement, territorial autonomy (i.e., removal from the market economy), "collective self-production of life," and individuality within the context of social relations (p. 387). Friedmann occasionally refers to particular places or examples, but he mostly develops his argument ahistorically. He gives few clues as to where and how his idealized community will come into existence.

By contrast, Patsy Healey concerns herself with actual planning practice (Healey, 1997). Thus, she avoids utopianism. At the same time, however, she indulges in a certain amount of sermonizing, and it is questionable whether her admonitions toward compromise and understanding constitute an effective strategy. She privileges the postmodernist value of diversity, along with the populist goal of participation, as her guiding norms, explicitly rejecting the political economy viewpoint. Thus, she dismisses what she calls the "pluralist socialist project," even when it respects difference, because it requires "a particular interpretation of what 'living together' and 'difference' mean . . . and thus cannot escape the critique of scientific rationalism" (Healey, 1996, p. 242). In other words, she is arguing that such a philosophy denies experiential knowledge by subordinating it to an abstract, general analysis of economic inequality.

Healey, like John Forester, builds on Jürgen Habermas's theory of communicative action (see Forester, 1989). Healey (1996) abandons any collective vision of the good city and instead falls back on a mediated, situational pluralism: " 'Right' and 'good' actions are those we can come to agree on, in particular times and places, across our diverse differences in material conditions and wants, moral perspectives, and expressive cultures and inclinations" (p. 243). Thus, she repudiates the scrutiny of outcomes characteristic of political-economy critiques and instead restricts herself to the evaluation of practices. Presumably, practices conducted in conformity with her model will produce good outcomes, at least according to the perceptions of the participants.

Healey (1996) herself notes the objection that social cleavage blocks consensus through debate but does not really answer it. Instead, she prescribes the "appropriate practices" of an "intercommunicative planning" (p. 246). Fundamental to her prescription is "respectful discussion within and between discursive communities" (p. 247). But this is not a realistic possibility within seriously divided societies. In the face of the deepening economic polarization of restructured capitalism, the call for respectful discussion is unlikely to provoke a meaningful response from the possessors of social power. Even when participants do verbally assent to a compromise and can agree on long-term goals, a negotiated consensus might fail to produce the desired outcome.

An illustration of the difficulty comes from South Africa, where a diverse group of stakeholders devised the post-apartheid government's housing policy:

> The loudly acclaimed "consensus" [on housing policy] supposedly hammered out by the stakeholders in the National Housing Forum which should have been achieved by hard bargaining among the parties was, in fact, the result of fudging vital differences between them. Faced with a conflict of vision between those who favored a market-oriented strategy led by the private sector and those who preferred a more "people-centered" approach in which "communities" would be the central players—or at least retain a veto—the forum parties opted for both, despite their incompatibility. Thus, all parties wanted immediate and visible delivery—but some also wanted "empowerment." So both were included, despite the fact that they would prove to be contradictory in practice.
>
> By the second year of implementation of the housing subsidy scheme, the consensus hammered out at the National Housing Forum had not, as its architects hoped, succeeded in binding all key housing interests to the policy; some key political actors had not been party to its formulation—and therefore did not feel bound by it—while crucial private interests proved ready to abandon it if it conflicted with their interests or if it did not seem to produce the rate of delivery that they had hoped to achieve. (Tomlinson, 1998, pp. 144-145)

The result of the unworkable consensus was inaction and failure even to approach the one goal on which everyone agreed—high levels of production.

Like Forester (1989), Healey relies on the works of Habermas for the philosophical basis of her prescriptions for city planning. The crux of Habermas's position lies in the command for truthfulness in communication, but in adopting this viewpoint, Healey shares in Habermas's evasion of the issue of social domination by capital. It is, in fact, puzzling that Habermas, rooted in the Marxian tradition, can propose truthful speech as the remedy for social evil without formulating a means for overcoming social structures that support distorted speech. Habermas (1971) argues that

> only in an emancipated society, whose members' autonomy and responsibility had been realized, would communication have developed into the nonauthoritarian and universally practiced dialogue from which both our model of reciprocally constituted ego identity and our idea of true consensus are always implicitly derived. (p. 314)

In a society that falls far short of the ideal of emancipation, undistorted speech might be desirable but in itself is not sufficient for combating hierarchically derived power.

Given the political weakness of the poor and marginalized, exhortations to mediating officials that they ensure transparency and a multiplicity of viewpoints respond only feebly to widespread social injustice. To some extent, Habermas originally was addressing Marxists in stressing his opposition to state bureaucratic domination. Now, however, the greatest threat to human welfare arises not from the vanquished red bureaucracies or the beleaguered administrators of local government but rather from the impersonal power of capital. Habermas's rediscovery of liberal principles of open discourse, albeit in more dialectical form, offers few improvements on John Stuart Mill's discussion of the marketplace of ideas. The classic criticisms of liberal pluralism point to the inability of relatively powerless groups to set the agenda for discussion and the unequal resources that different groups bring to the decision-making process (Bachrach & Baratz, 1962; Judge, 1995). These do not make liberalism's stress on rights and on the need to mediate among conflicting interests invalid; rather, they simply show it to be inadequate in the context of great inequalities of wealth and power (Katznelson, 1997). Unless the politically weak mobilize, they cannot insert themselves effectively into the debate, but strong mobilization rarely (if ever) comes out of the moderate, consensus-building rhetoric prescribed by proponents of the communicative turn. Bent Flyvbjerg comments, "Power knows that which Nietzsche calls 'the doctrine of Hamlet,' that is, the fact that often 'knowledge kills action; action requires the veils of illusion' " (Flyvbjerg, 1998, p. 229). Flyvbjerg is referring to the protective ideology of those in power, but his allusion applies equally to those pursuing empowerment. They must inspire fervent belief to overcome fear over the riskiness of oppositional action; reasoned counsel—undistorted speech—may indeed kill action. The language of rights, based on absolutist beliefs in the nature of a good society, offers a better vehicle for mobilization than does the language of compromise and mutual respect.

The work of Iris Marion Young, a philosopher who has attempted to reconcile the politics of difference with ethical precepts regarding justice, frequently is cited within the geography and planning literature. Young (1990) outlines a vision of the good city within the framework of a group-identified society. For Young, differences among groups give the city its character, whereas acceptance of difference provides the moral basis for urban life.[7] Such a stance fits well with certain depictions of the American city, for example, the "gorgeous mosaic" described by former New York Mayor David Dinkins or the picture drawn by Glazer and Moynihan (1970) in *Beyond the Melting Pot.* Even in the United States, however, the ideal of ethnic pluralism fails to come to grips with the endemic competition among identity groups for power and privilege. Common culture is not the only reason for group allegiance. Rather, group membership may allow privileged access to economic opportunity, and the desire to see members of one's group in a dominating economic position stems in part from the advantages that such ascendance gives other group members. The issue of group relations, therefore, is entangled with that of economic inequality.

In countries with stronger indigenous cultures than that of America, where the national culture is the primary identity group for most citizens, the adoption of diversity as a guiding value clashes with, rather than supports, the preservation of identity. The ideal of multiculturalism, although certainly opposed by many in the United States, is at least relevant in the framework of the American metropolis as well as in some other multiethnic locales and is itself part of the American mystique. On the other hand, in societies with "deep cultures," the notion of cultural diversity has little resonance. Furthermore, multiculturalism, which requires respect for the norms of others, offers few clues for dealing with those whose culture demands homogeneity.

In her discussion of the politics of difference, Young (1990) implies that urban division—what Katznelson (1981) calls city trenches—springs not from economic relations but rather from race, ethnicity, neighborhood, and gender. She rejects the old assimilationist ethic that provided the framework for absorption of immigrants during earlier times, arguing that it made unfair demands on the affected populations and that it was used for invidious discrimination. She favors explicit recognition of difference and the adoption of affirmative action procedures in school admissions and employment decisions.

Young (1990) agrees that no individual should be forced to accept a nominal group identity. Nevertheless, her view of the world leaves little opening for free choice of affiliation given that escape from an individual's primordial identities becomes highly problematic in an urban world without a dominant ideology of commonality. Even if the melting pot is a myth, it might be a desirable one, allowing individuals to transcend both internal and external labeling and providing a path to social peace through a sense of shared community. In a multiethnic urban society, where bonds already are tenuous, emphasizing difference can produce the ungovernable city, a milieu in which consensus is unachievable and in which forms of behavior emerge that others find intolerable. Thus, for example, Massey (1997) describes young men in a low-income housing estate who establish their identity by stealing cars for the fun of driving recklessly:

> Here was a clear establishment of an "us" and a "them" which demarcated both space-time and personal identity. . . . Such moves to establish exclusivity further debilitated any efforts to organize resistance to the deterioration of material and imaginary life [on the estate]. (p. 110)

Whether the outcome is a fertile disorder or the "wild city" (Castells, 1977), with each group hunkered down in its fortified enclave (Davis, 1990), depends crucially on whether group differences correlate with extreme economic differences. Thus, an ethos of diversity cannot be developed separately from an understanding of the economic bases of inequality.

In sum, although the theorists discussed in this section share a commitment to building a city that provides social justice and symbolic recognition to all its citizens, they differ on whether such a city is characterized by process or outcome, method or results. For Harvey, economic equality is fundamental, and negotiation and compromise represent a hopeless approach to its achievement. Friedmann's ideal community derives from the participation of its citizens in creating the conditions of their own existence. His is the picture of an end state, but not one that can be measured by income data. Healey's normative vision gives priority to the process through which decisions are made and stresses the significance of undistorted speech. Young's good city allows the retention of group identities and the explicit recognition of difference; her concern is to define a desirable set of social relations. The issue of whether any of these visions, or any concept of a good city, can be productive still must be addressed.

■ The Usefulness of Normative Analysis

Manuel Castells recently contested the assumption that explicit normative prescription is fundamental to coherent analysis and urban betterment. It is worth quoting at length Castells's (1998) argument, presented in the conclusion of his magisterial trilogy, *The Information Age.* Under the heading "What Is to Be Done?" he states,

> Each time an intellectual has tried to answer this question and seriously implement the answer, catastrophe has ensued. . . . Thus, . . . I shall abstain from suggesting any cure for the ills of our world. But . . . I would like to explain my abstention. . . .
>
> I have seen so much misled sacrifice, so many dead ends induced by ideology, and such horrors provoked by artificial paradises of dogmatic politics that I want to convey a salutary reaction against trying to frame political practice in accordance with social theory or, for that matter, with ideology. Theory and research, in general, . . . should be considered as a means for understanding our world and should be judged exclusively on their accuracy, rigor, and relevance. How these tools are used, and for what purpose, should be the exclusive prerogative of social actors themselves, in specific social contexts, and on behalf of their values and interests. (pp. 358-359)

Thus, Castells, who in *The Urban Question* began his career committed to class struggle and the triumph of urban social movements (Castells, 1977), now repudiates his youthful idealism.

Castells leaves behind as well the use of abstract systems of analysis; his approach is almost entirely inductive. Hence, his theoretical structure also differs from that of the other authors discussed in the preceding section, who proceed mainly deductively. Their arguments are derived from sets of premises and assumptions concerning governing values and operational possibilities (although Harvey and Healey do refer extensively to historical and practical examples in their writings). Castells, on the other hand, does not commit the error of unrealizable utopianism, nor does he expect that urging moral behavior is a particularly effective way in which to achieve it. Yet ultimately, his backing away from any formula for improvement is unsatisfactory. He leaves it to "social actors" to determine "their values and interests," but he offers few clues as to how they should go about doing so. Even Max Weber (the scope of whose work Castells [1998] is emulating in *The Information Age*), although deeply pessimistic about the future of European society, offers a moral code for political leadership and expresses a moderate faith in the usefulness of liberal institutions for promoting social well-being.

Castells deliberately offers us little assistance in answering the question of whether we can make the cities we want, and there is no shortage of failed projects in the history of the 20th century to support his stance. A continued belief that theoretical ideas can assist in producing a better world requires a resolute optimism, but such a stance offers the only route out of blind acceptance of the status quo. At any rate, despite Castells's dismissal of the objective of developing a normative framework, we can emulate his use of the inductive method. It provides a complementary approach to addressing normative questions, for any meaningful effort in this direction must work within the context of economic, social, and political forces. This means coming to grips with globalization, increased spatial and economic inequality, identity politics, and the contested role of the state as an agent of reform. It particularly means that arguments concerning planning and policy processes should not be isolated from analysis of the urban and regional system. We can begin with particular examples of success and failure in city building and ask whether these cases give rise to general lessons, although such lessons still must be examined in the light of abstract categories. Like any search for best practices, identifying successful cities raises questions of the transferability of findings. Is it the practice itself or the particular history of a place and the character and effectiveness of its leadership that produce desirable outcomes? Thus, the search for urban models is not sufficient without also including a logical analysis of their general applicability. Empiricism, moreover, cannot transcend what already exists. Nevertheless, it at least allows a measure of the flexibility available for social restructuring within the constraints of world capitalism.

Planning and Policy in Two Global Cities

Examination of particular cities to discover the values incorporated within their development offers one approach to discovering whether or not we can make the cities we want. If examples exist of urban milieus that combine equality, diversity, participation, and sustainability, then we can argue that these are more than simply abstract values. In addition, we can attempt to discover the sources of their moral structure. Given our particular concerns with issues of diversity and equality, a comparison of Amsterdam and New York is useful in showing the possibility of embedding these values within the context of large, complex urban places.

Amsterdam and New York share certain characteristics despite New York's greater size and economy. New York is incontestably a global city; Amsterdam, when considered as part of the *Randstaat,* can be regarded as similar, if not in the same league. Both are financial centers and have endured the recent restructuring of their economies, with massive decline of manufacturing and port facilities and jumps in the size of their service (especially business service) sectors. Both have ethnically diverse populations and have absorbed major flows of immigration within the past 30 years, although Amsterdam differs from New York in the tapering off of that flow. Although each can make claims to embodying the values of diversity and participation, the two cities differ sharply in terms of equality.

Amsterdam

Amsterdam offers the best available model of a relatively egalitarian, diverse, democratic city with a strong commitment to environmental preservation.[8] Its geographic situation, presenting the constant threat of inundation and requiring construction on land below sea level, means that urban development has, for centuries, proceeded as the result of conscious decisions. Within a country possessing a long tradition of bourgeois paternalism and a more recent one of working-class organization and social democratic leadership, Amsterdam deviates from the rest of the Netherlands in its greater population diversity and more extensive range of lifestyles. In this respect, Amsterdam also differs from other relatively egalitarian northern European cities:

> The most extraordinary quality of this city [is] its relatively . . . successful achievement of highly regulated urban anarchism, another of the paradoxes . . . that two-sidedly filter through the city's historical geography in ways that defy comparison with almost any other *polis,* past or present. This deep and enduring commitment to libertarian socialist values and participatory spatial democracy is apparent throughout the urban built environment and in the social practices of urban planning, law enforcement, popular culture, and everyday life. (Soja, 1996, p. 285)

During the 19th century, the Dutch reached a compromise between their antagonistic Protestant and Catholic factions. The form of this consensus, termed "pillarization," involves separate sets of civic institutions, including state-subsidized schools, based in the different religious communities. These religiously based institutions continue to act as important intermediaries between the Dutch state and recipients of social services and social housing. The Dutch tradition of philanthropy and relative freedom from bureaucratic constraint makes the Netherlands' welfare system more flexible than most of its European counterparts.

During the 20th century, working class mobilization and trade union militancy in the Netherlands produced a typical European multiparty system with coalition governments and socialist representation. After World War II, demands for income security and better living conditions were accommodated through corporatist negotiation among unions, industry, and the state. A system of accommodation based on class politics, however, proved inadequate during the 1970s and 1980s in the face of urban social movements. These movements, especially as embodied in large-scale squatting, broad-based resistance to the eviction of squatters, and popular opposition to urban renewal, caused the Amsterdam municipal government to retreat from the top-down planning methods that had been acceptable to traditional socialists.

Now, Amsterdam's government incorporates the bourgeoisie's historic commitment to charity, the socialist concern with inequality, and the participatory thrust of the post-1960s urban rebellion. The overlay of struggle and compromise has offered a framework in which contemporary difference is negotiated. Amsterdammers, in line with Healey's criterion for urban tolerance, live together but differently. They do so within a city remarkably integrated spatially by class, less so by ethnicity (Musterd, Ostendorf, & Breebart, 1998). The state, both national and local, relies heavily on intermediaries, rooted in different communal groupings, as agents for the dispensation of social welfare. Although religion has faded in importance as a fault line in Amsterdam society, religiously based institutions continue to play a significant role. The philanthropic tradition that derives from them combines with leftist political commitment to create a milieu resistant to marketization. By and large, the upper strata of Amsterdam society accept high levels of redistribution, mixed neighborhoods, and cramped housing to experience the benefits of a cosmopolitan, egalitarian city.

The willingness of Amsterdam's citizens to tolerate their neighbors is part of a virtuous circle resulting from a generous welfare state. No severely impoverished lower class threatens social peace, and even the most ghettoized areas of the city are neither homogeneously low income nor populated entirely by immigrant groups. Consequently, difference is less frightening than in other locales, where it is more strongly associated with crime and danger. Tolerance of the use of soft drugs, medicalization of addiction, and police intervention to prevent hard drug sales from getting out of control inhibit the association between criminality and poor communities that prevails elsewhere. It is hard to imagine that similar conditions could exist if Amsterdam had more extreme income inequality.

The situation of Amsterdam bolsters the case of those who argue that local autonomy is possible and that globalization does not destroy the capacity of nation-states to achieve social goals. Although Amsterdam depends heavily on the Dutch national state to maintain income support for its inhabitants, local decisions on housing policy, including the legalization of squatting and of berthing of houseboats in central locations, contribute markedly to the city's heterogeneity. The types of speculative activity that have produced rapid gentrification in many European cities are largely absent here. Residential property developers are willing to accept planning and the security of assured profits instead of riskier, if potentially larger, speculative gains. Amsterdam, in many respects, does incorporate the values of equality, diversity, democracy, and respect for the environment.

Although there now are moves toward reduction in the welfare state and privatization, they are limited.[9] Amsterdam, of course, is not a utopia. It does display income inequality and ethnic segregation. Much of its postwar housing stock, which is occupied primarily by Turks and Moroccans, is well below contemporary standards and is deteriorating, and few funds are earmarked for its upgrading. The substantial housing rights of sitting tenants make it very difficult for newcomers to enter the city. The Green Heart, the large area of open space that forms the core of the Randstaat and, along with the dominance of bicycle transport within central Amsterdam, represents the city's environmental values, is suffering from encroachments of various sorts. Nonetheless, relative to other cities, Amsterdam does strongly embody progressive values.

The explanation for its success lies in the intertwining of elements of politics, civic action, ideology, and planning. Whether this amalgam is transferable elsewhere and can be used to demonstrate the possibility of attaining the city we want is a vexing question. Indeed, whether Amsterdam itself can continue to sustain its current commitments in the face of middle class demands for centrally located housing, the withdrawal of the central government from subsidizing new construction, the forces of international competition, and developmental pressures on open space remains to be seen.

New York City

New York offers a troubling contrast to Amsterdam. Despite an economic boom, income inequality worsens, basic infrastructure deteriorates, the housing problems of low-income people grow, school buildings crumble while class sizes increase, and a steady stream of incidents involving the police and people of color mark the racial divide. At the same time, crime has dropped dramatically, survey results show New Yorkers feeling that their city is improving, and the city is home to an astonishing number of new immigrants.[10]

Historically, New York pioneered social legislation. At the beginning of the 20th century, it introduced tenement laws, and its active labor movement successfully pressed for wages and hours legislation. After World War II, massive subsidies for working and middle class dwellings and rent regulation kept housing affordable for millions of citizens. The City University offered free tuition, and the extensive municipal hospital system provided health care for the uninsured. Still, the suburbanization of much of the upper and middle class, the disciplining of the city's government that followed the 1975 fiscal crisis, and the decline of the white industrial working class ended the possibility that the city by itself could or would play a strong role in promoting economic equality.

New York demonstrates that ethnic toleration can be quite compatible with economic inequality and, in fact, can foster it. Despite continued occurrences of racial and ethnic hostility, new immigrants, mainly of color, have found jobs and revived neighborhoods (Winnick, 1990) during the past two decades. Its otherwise conservative mayor, Rudolph Giuliani, has welcomed newcomers; unlike their Los Angeles counterparts, New York politicians rarely use nativist sentiments as a basis for electoral appeal. Although residential segregation persists, New York's public spaces display a remarkable mixture of physical types, and except in certain ethnic enclaves, interracial relationships provoke little hostility. The city's public bureaucracy disproportionately incorporates African Americans, and the school system celebrates the rainbow of identities represented in it.

New York shows the limits of diversity and tolerance of "otherness" as a basis for social reconstruction. Extreme income inequality and the persistence of large, racially defined areas of great deprivation coexist with verbal recognition of multiculturalism and a highly absorptive political system. Words, however, cannot overcome inequalities in power that are fundamentally rooted in access to resources, even though a public rhetoric of exclusion and vilification obviously would make matters worse.[11]

Just as Amsterdam's relative egalitarianism depends on the redistributive character of the Dutch state, New York's racial and income inequities incorporate the nature of its national state. The United States lacks both the philanthropic religious tradition and the left mobilization that shaped the Dutch welfare system. Thus, the practices of Amsterdam, dependent as they are on the Dutch context, are not easily emulated across the Atlantic.

■ Can Something Be Done, and What Should It Be?

The American sociopolitical context, with its absence of a strong welfare state and the ontological security that such a state creates, forestalls the possibility that cities will combine diversity, participation, and equity. The Amsterdam case implies that democratic procedures and just actions flow from situations where social citizenship already exists (cf. Marshall, 1965). It is possible to develop a view of social justice that transcends a particular social context, but its implementation requires that elements of realization already be present. To put this another way, achievement of the city we want is a circular process, whereby the preexistence of equity begets sentiments in its favor, democratic habits produce popular participation, and diversity increases tolerance.

Thus, substance and procedure are inseparable. The just city incorporates just outcomes, not simply open processes: "Political and economic social relationships are intrinsically intertwined and cannot be torn asunder" (Lauria, 1997, p. 41). Substance and procedure must be contemplated simultaneously; likewise, desirable end states must be considered in tandem with the forces to achieve them. Thus, movement toward a more desirable city presents baffling strategic problems because mobilizing a force sufficient to overcome barriers to change demands a messianism that contravenes undistorted speech and can provoke fierce reaction.

If Amsterdam presents a rough image of a desirable urban model, then strategies and normative emphases will differ in respect to reaching that goal, depending on the starting point. Despite the famous Dutch predilection for arriving at decisions by compromise, the preconditions for consensus were the development of strong oppositional forces including a class-conscious labor movement and a militant, sometimes violent, squatters movement. Amsterdam does illustrate the potential of decision-making through negotiation, but only within a history of preexisting mobilizations. Now, as those mobilizations have subsided, an erosion of state-sponsored redistributive measures has begun. If the result is a major decline in economic equality, then the basis by which consensual politics contributes to a self-reinforcing cycle of tolerance and equity will ebb.

In American cities, practical strategies for the development of a more just and tolerant system require the dampening of sentiments based on group identity, greater commitment to common ends, and identification of institutions and policies that offer broadly appealing benefits. The most effective approach uses cross-cutting group identities to bolster unity. During the 1960s, successful movements were based on groups that shared a common identity because of similar racial, territorial, and client statuses (Fainstein & Fainstein, 1974). In the new century, common identity must be defined differently. Effectiveness requires organizing around work status when it overlaps with immigrant or gender situation. It also means continued organizing around affordable housing, but this must include recognition of the housing needs of the middle class, not simply a call for assistance to the poorest. Narrowly targeted policies, however efficient in producing redistribution or righting past wrongs, lack a sufficient constituency and seem unjust to those who do not benefit from them (Sayer & Storper, 1997b).

Whereas the urban social movements of the past centered on collective consumption organized around residence, future movements need to address the organization of work and must concern themselves with the consumption issues of new types of workers. The changing nature of work calls for unions of temporary workers, household workers, and the self-employed; rather than organizing around the workplace, this would require organizing around the type of work conducted. Such unions would have to emphasize their service role—job training and placement, establishment of benefit pools and portability of benefits, provision of legal services, credit unions, mortgage assistance. The model is craft rather than industrial unionism.

In most cities on the European continent, absolute need on the American scale is not present. Especially in France and Germany, with traditions of a strong state and strong pressures for homogeneity and immigrant exclusion, the plea for citizen participation, negotiation, and a less authoritative government is progressive. Because the sense of the collectivity already is present, a more transactional, participatory approach represents reform. In the United States, where most cities are dominated by business-led regimes (Stone, 1993), individual citizen participation will not provide a path to social transformation, even though it might produce better outcomes in particular cases. Urban citizen participation, as it is conducted now, mainly involves participants demanding marginal changes in the status quo or benefits that respond to their narrowly defined interests.

Movement toward a normative vision of the city requires the development of counterinstitutions capable of reframing issues in broad terms and of mobilizing organizational and financial resources to fight for their aims. As is, in the United States, no broad-based media exist to communicate alternative approaches to questions raised by urban economic development, metropolitan inequalities, and environmental preservation. The inherently divisive character of identity politics cuts against the building of such institutions and, therefore, can only be self-defeating. As Neil Smith contends, demands based on group identity initially countered possessive individualism but ultimately undercut the movement toward greater social justice. By provoking those already in relatively privileged positions to demand their rights, identity politics "becomes a vehicle for a reassertion of individualism—a hallmark of the revanchist city" (N. Smith, 1997, p. 134). In such a context, there is no shared vision of a desired future.

Smith is dubious that any abstract conception of justice can, or even should, provide such a vision. Without one, however, we have no means of persuading people that they should follow any guidance beyond their own narrow self-interests or that anything can be gained from the collective enterprise. Amsterdam's achievements are not simply a response to a militant working class. Rather, they also depend on a widely felt sense of justice. Enough of the upper social strata accept a moral code that they do not resist, and even support, redistributional measures.

Concepts of justice can be relational to context (i.e., not purely abstract) without being wholly relative (D. M. Smith, 1997). We need to place ourselves on this tenuous middle ground of being able to compare, of being able to say what is better and what is worse even if we cannot say what is good and what is bad, if we are to develop the cities we want.

NOTES

1. Sayer and Storper (1997b) comment, "Any social science claiming to be critical must have a standpoint from which its critique is made. . . . It is strange that this [recent] critical social science largely neglects to acknowledge and justify these standpoints" (p. 1). See also Fainstein (1997b).

2. See Beauregard (1989), Fischer and Forester (1993), Harvey (1996), Healey (1996, 1997), Merrifield and Swydegouw (1997), Sayer and Storper (1997a), and Sennett (1990).

3. Many would not accept my premise. To enter the debate over the usefulness of planning, however, would constitute too lengthy a digression from the main theme of this chapter. See Fainstein and Fainstein (1979) and Klosterman (1985).

4. See the critiques by Castells (1977) and Logan and Molotch (1987).

5. See, inter alia, Castells (1977), Fainstein and Fainstein (1979), Foglesong (1986), and Harvey (1989b).

6. Similarly, Giddens (1990) refers to his vision of a postmodernist world (by which he means a quite different concept from the usual definition) as "utopian realism."

7. See Jacobs (1961) and Sennett (1970) for similar moral stances.

8. The discussion of Amsterdam is drawn from Fainstein (1997a). That article was based on research conducted while I was a visiting professor at the University of Amsterdam within the framework of the *Wibautchair,* sponsored by the city of Amsterdam.

9. For example, the Dutch government has largely retreated from the construction of new social housing. At the same time, it has substantially increased housing allowances, which are available to all those with incomes below a certain level (Priemus, 1998).

10. Information on New York City is drawn from Fainstein (1998) and Fainstein and Fainstein (1978, 1989).

11. Butler (1997) explains why speech injures despite its immateriality. The argument that New York's public discourse is one of tolerance is not intended to disguise the reality of racial and ethnic tension, politicians who pander to racist sentiment, or official callousness toward immigrants.

REFERENCES

Altshuler, A. A. (1965). *The city planning process.* Ithaca, NY: Cornell University Press.

Anderson, M. (1964). *The federal bulldozer.* Cambridge, MA: MIT Press.

Bachrach, P., & Baratz, M. S. (1962). Two faces of power. *American Political Science Review, 56,* 947-952.

Beauregard, R. A. (1989). Between modernity and postmodernity: The ambiguous position of U.S. planning. *Environment and Planning D: Society and Space, 7,* 381-395.

Berman, M. (1988). *All that is solid melts into air.* Harmondsworth, UK: Penguin.

Butler, J. (1997). *Excitable speech.* London: Routledge.

Castells, M. (1977). *The urban question.* Cambridge, MA: MIT Press.

Castells, M. (1998). *The information age: Economy, society, and culture: Vol. 3. End of millenium.* Oxford, UK: Blackwell.

Davidoff, P. (1965). Advocacy and pluralism in planning. *Journal of the American Institute of Planners, 31,* 544-555.

Davis, M. (1990). *City of quartz.* New York: Vintage.

Fainstein, S. S. (1997a). The egalitarian city: The restructuring of Amsterdam. *International Planning Studies, 2*(3), 295-314.

Fainstein, S. S. (1997b). Justice, politics, and the creation of urban space. In A. Merrifield & E. Swyngedouw (Eds.), *The urbanization of injustice* (pp. 18-44). New York: New York University Press.

Fainstein, S. S. (1998). Assimilation and exclusion in U.S. cities: The treatment of African Americans and immigrants. In S. Musterd (Ed.), *Urban segregation and the welfare state* (pp. 28-44). London: Routledge.

Fainstein, S. S., & Fainstein, N. (1974). *Urban political movements.* Englewood Cliffs, NJ: Prentice Hall.

Fainstein, S. S., & Fainstein, N. (1978). National policy and urban development. *Social Problems, 26,* 125-146.

Fainstein, S. S., & Fainstein, N. (1979). New debates in urban planning: The impact of Marxist theory within the United States. *International Journal of Urban and Regional Research, 3,* 381-403.

Fainstein, S. S., & Fainstein, N. (1989). Governing regimes and the political economy of redevelopment in New York City. In J. Mollenkopf, T. Bender, & I. Katznelson (Eds.), *Power, culture, and place: Essays on the history of New York City* (pp. 161-199). New York: Russell Sage.

Fainstein, S. S., & Fainstein, N. (1996). City planning and political values: An updated view. In S. Campbell & S. S. Fainstein (Eds.), *Readings in planning theory* (pp. 265-287). Oxford, UK: Blackwell.

Fischer, F. (1990). *Technocracy and the politics of expertise.* Newbury Park, CA: Sage.

Fischer, F., & Forester, J. (Eds.). (1993). *The argumentative turn in policy analysis and planning.* Durham, NC: Duke University Press.

Fishman, R. (1977). *Urban utopias in the twentieth century.* Cambridge, MA: MIT Press.

Flyvbjerg, B. (1998). *Rationality and power.* Chicago: University of Chicago Press.

Foglesong, R. E. (1986). *Planning the capitalist city.* Princeton, NJ: Princeton University Press.

Forester, J. (1989). *Planning in the face of power.* Berkeley: University of California Press.

Friedmann, J. (1987). *Planning in the public domain.* Princeton, NJ: Princeton University Press.

Gans, H. (1968). *People and plans.* New York: Basic Books.

Giddens, A. (1990). *The consequences of modernity.* Stanford, CA: Stanford University Press.

Glazer, N., & Moynihan, D. P. (1970). *Beyond the melting pot* (2nd ed.). Cambridge, MA: MIT Press.

Habermas, J. (1971). *Knowledge and human interests.* Boston: Beacon.

Hall, P. (1988). *Cities of tomorrow.* Oxford, UK: Basil Blackwell.

Harvey, D. (1985). On planning the ideology of planning. In *The urbanization of capital* (pp. 165-184). Baltimore, MD: Johns Hopkins University Press.

Harvey, D. (1989a). *The condition of postmodernity.* Oxford, UK: Blackwell.

Harvey, D. (1989b). *The urban experience.* Baltimore, MD: Johns Hopkins University Press.

Harvey, D. (1996). *Justice, nature, and the geography of difference.* Oxford, UK: Blackwell.

Hayek, F. (1944). *The road to serfdom.* Chicago: University of Chicago Press.

Healey, P. (1996). Planning through debate: The communicative turn in planning theory. In S. Campbell & S. S. Fainstein (Eds.), *Readings in planning theory* (pp. 234-257). Oxford, UK: Blackwell.

Healey, P. (1997). *Collaborative planning: Shaping places in fragmented societies.* Vancouver, British Columbia: UBC Press.

Jacobs, J. (1961). *Death and life of great American cities.* New York: Vintage.

Judge, D. (1995). Pluralism. In D. Judge, G. Stoker, & H. Wolman (Eds.), *Theories of urban politics* (pp. 13-34). London: Sage.

Katznelson, I. (1981). *City trenches.* New York: Pantheon.

Katznelson, I. (1997). Social justice, liberalism, and the city: Considerations on David Harvey, John Rawls, and Karl Polanyi. In A. Merrifield & E. Swyngedouw (Eds.), *The urbanization of injustice* (pp. 45-64). New York: New York University Press.

Klosterman, R. E. (1985). Arguments for and against planning. *Town Planning Review, 56*(1), 5-20.

Lauria, M. (Ed.). (1996). *Reconstructing urban regime theory: Regulating urban politics in a global economy.* Thousand Oaks, CA: Sage.

Lauria, M. (1997, Summer). Communicating in a vacuum: Will anyone hear? *Planning Theory,* pp. 40-42.

Lindblom, C. E. (1959). The science of muddling through. *Public Administration Review, 19,* 79-88.

Logan, J., & Molotch, H. (1987). *Urban fortunes.* Berkeley: University of California Press.

Lucy, W. H. (1988). APA's ethical principles include simplistic planning theories. *Journal of the American Planning Association, 54,* 147-149.

Marshall, T. H. (1965). *Class, citizenship, and social development.* New York: Anchor.

Massey, D. (1997). Space/power, identity/difference: Tensions in the city. In A. Merrifield & E. Swyngedouw (Eds.), *The urbanization of injustice* (pp. 100-116). New York: New York University Press.

Merrifield, A., & Swydegouw, E. (Eds.). (1997). *The urbanization of injustice.* New York: New York University Press.

Musterd, S., Ostendorf, W., & Breebart, M. (1998). *Multi-ethnic metropolis.* Dordrecht, The Netherlands: Kluwer.

Needleman, M. L., & Needleman, C. (1974). *Guerrillas in the bureaucracy.* New York: John Wiley.

Priemus, H. (1998). Improving or endangering housing policies? Recent changes in the Dutch housing allowance scheme. *International Journal of Urban and Regional Research, 22,* 319-330.

Sayer, A., & Storper, M. (Eds.). (1997a). Ethics unbound [special issue]. *Environment and Planning D: Society and Space, 15*(1).

Sayer, A., & Storper, M. (1997b). Ethics unbound: For a normative turn in social theory. *Environment and Planning D: Society and Space, 15*(1), 1-18.

Sennett, R. (1970). *The uses of disorder.* New York: Vintage.

Sennett, R. (1990). *The conscience of the eye.* New York: Knopf.

Smith, D. M. (1997). Back to the good life: Towards an enlarged conception of social justice. *Environment and Planning D: Society and Space, 15*(1), 19-36.

Smith, N. (1997). Social justice and the new American urbanism: The revanchist city. In A. Merrifield & E. Swyngedouw (Eds.), *The urbanization of injustice* (pp. 117-136). New York: New York University Press.

Soja, E. W. (1996). *Thirdspace.* Oxford, UK: Blackwell.

Squires, G. (Ed.). (1989). *Unequal partnerships.* New Brunswick, NJ: Rutgers University Press.

Stone, C. N. (1993). Urban regimes and the capacity to govern: A political economy approach. *Journal of Urban Affairs, 15,* 1-28.

Toll, S. (1969). *Zoned American.* New York: Grossman.

Tomlinson, M. R. (1998). South Africa's new housing policy: An assessment of the first two years, 1994-96. *International Journal of Urban and Regional Research, 22*(1), 137-146.

Winnick, L. (1990). *New people in old neighborhoods.* New York: Russell Sage.

Young, I. M. (1990). *Justice and the politics of difference.* Princeton, NJ: Princeton University Press.

13

The Spaces of Democracy

RICHARD SENNETT

About 20 years ago, I went to Jerusalem as part of a planning group from the Graduate School of Design at Harvard University. Although we knew better rationally, we were fired up by the belief that art might succeed in making a more democratic city in which politics had failed. My team explored how to transform a triangle of wasteland outside the Damascus Gate into a public space that Palestinians and Israelis might share. The team assigned me the task of meeting with Palestinian officials because I previously had done planning work in Jordan; specifically, I was asked to enlist the help of Anwar Nusseibeh, the doyen of an old and elite Palestinian family.

I first went to visit Nusseibeh at his office. He headed the East Jerusalem Electric Light Company, one of the few local businesses the Israelis allowed Palestinians to manage in the city. Nusseibeh was a courtly man who, I discovered, in a better life would have devoted himself to poetry rather than to electricity. By chance, we had slipped into speaking French at that first encounter, and he began to describe to me writers and artists he had known in Paris during the 1930s; these figures were more alive in his memory than were the immediate difficulties he faced.

I cannot say that a bond of trust developed between us because I could do nothing about being American and a good Jew or about the driver and guards provided by the mayor of Jerusalem. As the afternoon light faded in his office, while we spoke in a language foreign to the Israeli monitors, we began to understand one another. Our talks continued over the next few days in a cafe and finally in Nusseibeh's home, mostly about prewar Paris. France had been, for different reasons, a refuge for each of us; something about Paris arouses in its foreign residents feelings of regret for the past. In any event, this shared bond prompted Nusseibeh, the most courteous of men, finally to challenge me about the present. Nusseibeh said (paraphrasing from my notes),

> You want to build a place at Damascus Gate for "democracy," but you cannot show me—even supposing democracy is possible between victors and the people they have captured—what a democratic space looks like. Will better buildings incline the Israeli people to treat us as equals, better buildings curb the violent rage of our own young ones? As I say, even if we forget our impossible present circumstances, what effect can the mere shape of a wall, the curve of a street, lights and plants, have in weakening the grip of power or shaping the desire for justice?

Nusseibeh took the occasion of the ensuing silence to pour me more tea.

Nusseibeh's challenge to me had two mental parts: how visual design might serve the political project of democracy and what is urban about democracy itself. The context of our discussion was not a philosophy seminar. Moreover, the challenge haunts urban designers today in places as diverse as Sarejevo after its civil war, Berlin after the fall of communism, and Los Angeles after its racial riots. In these places, as in Jerusalem, the politics of conflict is hard to relate to urban design. Yet, the essence of democracy lies in displacing conflict and difference from the realm of violence to a more peaceable, deliberative realm. How to do so was Nusseibeh's challenge to me.

■ ■ ■

Jerusalem is a very old city, and in ancient times, those who lived in Jerusalem might have known how to respond to Nusseibeh's challenge by invoking examples from Athens, the center of civilization in the ancient world. From roughly 600 to 350 BC, Athens located its democratic practices in two places in the city: the town square and the theater. Two very different types of democracy were practiced in the square and the theater. The square stimulated citizens to step outside their own concerns and note the presence and needs of other people in the city. The architecture of the theater helped citizens to focus their attention and concentrate when engaged in decision making.

We never would want to copy the social conditions of Athenian democracy. The majority of people living in the city were slaves, and all women were excluded from politics. Still, we can learn something from how this often fickle, intensely competitive people related democracy such as they knew it to architecture.

It was in the Pnyx that the Athenians debated and decided on the actions the city would take. The Pnyx was a bowl-shaped, open-air theater about a 10-minute walk from the central square of Athens. Chiseled out of a hill, the Pnyx resembled in form other Greek theaters and, like them, originally provided space for dancing and plays. In the sixth and fifth centuries BC, Athenians put this ordinary theater to a different use in seeking order in their politics. The speaker stood in the open round space on a stone platform called a *bema* so that he or she could be seen by everyone in the theater. Behind the speaker, the land dropped away so that words seemed to hover in the air between the mass of 5,000 to 6,000 bodies gathered together and the empty sky. The sun from morning to late afternoon struck the speaker's face so that nothing in his expression or gestures was obscured by shadow. The audience for this political theater sat around the bowl in assigned places, men sitting with others who belonged to the same local tribe. The citizens watched each other's reactions as intently as they did the orator at the bema.

People sat or stood in this relation for a long time—so long as the sunlight lasted. Thus, the theatrical space functioned as a detection mechanism, its focus and duration meant to get beneath the surface of momentary impressions. Such a disciplinary space of eye, voice, and body had one great virtue: Through concentration of attention on a speaker and identification of others in the audience who might call out challenges or comments, the ancient political theater sought to hold citizens responsible for their words.

In the Pnyx, two visual rules organized the often raucous meetings at which people took decisions: exposure, both of the speaker and of the audience to one another, and fixity of place, in terms of where the speaker stood and the audience sat. These two visual rules supported a verbal order—a single voice speaking at any one time.

The other space of democracy was the Athenian agora. The town square consisted of a large open space crossed diagonally by the main street of Athens. At the sides of the agora were temples and buildings called *stoas,* the latter being sheds with open sides onto the agora. A number of activities occurred simultaneously in the agora—commerce, religious rituals, casual hanging out. In the open space also lay a rectangular law court, surrounded by a low wall, so that citizens banking or making offerings to the gods also could follow the progress of justice. The stoa helped to resolve this confusion; as one moved out of the open space inside the building, one moved from a public realm in which citizens freely intermingled into more private spaces. The rooms at the back of the stoas were used for dinner parties and private meetings. Perhaps the most interesting feature of the stoa was the transition space just under the shelter of the roof on the open side; here, one could retreat yet keep in touch with the square and its activities.

What import did such a complex, teeming space have for the practice of democracy? A democracy supposes that people can consider views other than their own. This was Aristotle's notion in the *Politics.* He thought that the awareness of difference occurs only in cities because every city is formed by *synoikismos,* a drawing together of different families and tribes, of competing economic interests, of natives with foreigners.

"Difference" today seems about identity; we think of race, gender, or class. Aristotle meant something more by difference; he included the experience of doing different things, of acting in divergent ways that do not neatly fit together. The mixture in a city of action as well as identity is the foundation of its distinctive politics. Aristotle's hope was that when a person becomes accustomed to a diverse, complex milieu, he or she will cease reacting violently when challenged by something strange or contrary. Instead, this environment should create an outlook favorable to discussion of differing views or conflicting interests. The agora was the place in the city where this outlook should be formed.

Nearly all modern urban planners subscribe to this Aristotelian principle. Diversity loses its force, however, if in the same space different persons or activities are merely concentrated but each remains isolated and segregated. Differences have to interact.

The Athenian agora made differences interact among male citizens in two ways. First, in the open space of the agora, there were few visual barriers between events occurring at the same time so that men did not experience physical compartmentalization. As a result, in coming to the town square to deal with a banker, one might suddenly be caught up in a trial occurring in the law court, shouting out one's own opinion or simply taking in an unexpected problem. Second, the agora established a space for stepping back from engagement. This occurred at the edge, just under the roof of the stoa on its open side; here was a fluid, liminal zone of transition between private and public.

These two principles of visual design, lack of visual barriers but a well-defined zone of transition between public and private, shaped people's experience of language. The flow of speech was less continuous and singular than in the Pnyx; in the agora, communication through words became more fragmentary as people moved from one scene to another. The operations of the eye were correspondingly more active and varied in the agora than in the Pnyx; a person standing under the stoa roof looked out, his or her eye searching and scanning. In the Pnyx, the eye was fixed on a single scene, that of the orator standing at the bema; at most, the observer scanned the reactions of people sitting elsewhere, fixed in their seats.

■ ■ ■

This ancient example illustrates how theaters and town squares can be put to democratic use. The theater organizes the sustained attention required for decision making; the square is a school for the often fragmentary, confusing experience of diversity. The square prepares people for debate; the theater visually disciplines their debating.

This is, of course, in principle. Throughout their long history, these two urban forms have been put to many divergent or contrary uses. We need only think of the Nazi spectacles in Germany to summon an image of theatrically focused attention dedicated to totalitarian ends, and the disorders of 19th-century Parisian squares frequently drove people further inside themselves rather than making citizens more attentive to each other.

Yet, the mind creates by considering models, ideals, and possibilities. For me, at the time I went to Jerusalem, the model of the agora was the touchstone of my love for cities and in my faith in urban design, as it was for other urbanists such as Jane Jacobs and Henri Lefebvre and, more largely, for radicals of the 1960s. I knew one big thing when I began to write. Every individual needs the experience of being challenged by others to grow both psychologically and ethically. Psychologically, human beings develop only in a rhythm of disorientation and recovery; a static sense of self and world becomes a type of psychological death. Ethically, painful and uncomfortable encounters with those who differ are the only ways in which individuals learn modesty. For these reasons, I believed, human beings need cities, and within cities agoras of some sort, to become fully human.

I could have summoned these arguments when Nusseibeh challenged me about plans for the Damascus Gate. I had reasoned them through in my first book, *The Uses of Disorder,* and spent a decade thereafter trying to realize them in practice. But I remained silent. In looking back, I understand the reasons why I said nothing. First, I would have answered him in bad faith, as an American urbanist speaking about democracy. Second, in Jerusalem, I began to lose my faith in the agora.

■ ■ ■

A future historian might well conclude that Americans during the last half of the 20th century focused their energies on preventing democracy in the built environment. Gated communities, now the most popular form of American residential building, take to an extreme denial of democracy of the agora sort; here are homogenized communities guarded and sealed off like medieval castles. In my youth, less extreme forms of American development already tended to the same end. The shopping malls of the 1920s through 1950s were indeed diverse places. The malls that came into existence during the 1960s were monofunctional; today, one rarely will see an AIDS service agency or a police station in a mall or a Gap store next door to a school. Moreover, the renewal of old cities like my own, New York, had depended on the globalization of the world economy. Globalization creates cities that are sharply divided, and a globalized core now isolates Manhattan, for example, from the localized economies and cultures of the city's outer boroughs.

Professional urban design is part of this story of bad faith with democracy. The pristine, white-gleaming small towns produced by the movement called the New Urbanism are a world apart from the everyday disorders of life; the kitsch, pseudo-small towns now being built as an antidote to suburban sprawl provided no home for differences—differences of the sort that lead to conflicts of ethnicity, race, class, and/or sexual preference. In a purely stylistic vein, the battle between modernism and postmodernism is a clothing conflict about the surfaces of buildings, and these outer architectural garments tell us little about how to make buildings and spaces more democratic.

It could be said that the American city only reflects larger currents of American culture. American culture has indeed put a premium on difference in the "identity talk" that emphasizes distinctions, particularly between that familiar friend we love to hate—the white, middle class heterosexual, Anglo-Saxon male—and all those he has (at least in theory) oppressed.

Identity talk of the American sort leads to isolation rather than to interaction. Our culture prefers clear pictures of self and social context. For the sake of this clarity, and for the sake of identity, we sacrifice democracy—democracy in Aristotle's sense of the dialogues, debates, and shared deliberations that might take us out of ourselves and the sphere of our immediate self-knowledge and interests. Writers from other cultures urge us to break out of identity ghettoes. Stuart Hall does so in his writings on the hybrid identities of people who move geographically or socially, and Homi Bhabba contests the ghetto of the self by exploring the positive aspects of uncertainty when a person is in the presence of an alien "other." Still, these writers have not found a general public in their adopted country.

■ ■ ■

In 1980, when I went to Jerusalem, American ways of denying the agora were partly why I fell silent when challenged by Nusseibeh, an admission that I had come to him empty-handed. This he accepted in good grace by dropping the distasteful subject and returning tactfully to the origins of surrealism in Paris.

Yet, going to Jerusalem was an important event for me as an urbanist. The city challenged my belief in the agora, at least as school for democracy. Jerusalem's old city within the walls is filled with the human differences that thousands of years of conquest, migration, faith, and trade have laid on the land like a thick impasto on canvas. In its covered shopping streets, Jews and Muslim shopkeepers mix together in pursuit of trade and tourists. On the via Dolorosa, the processions of Christian pilgrims stream past the small shops of nonbelievers who acknowledge the pilgrims' faith by leaving the pilgrims alone in silence. When the right-wing Israeli government has sought to dig beneath the holy Islamic shrine of al-Aksa, many Jewish residents in the city have turned out in protest. All these are signs of the living presence of the agora.

Still, Jerusalem is hardly at peace. The spirit of the agora permeated Sarajevo before the civil war or, in a more moderate fashion, exists in post-Communist Berlin. All these places have known daily and painful encounters with difference, yet the encounters alone have not bred civic bonds. If these cities have various modern versions of the agora, then they lack any effective equivalent of a Pnyx. I do not mean to suggest that I suddenly stopped believing in the value of living in difference, only that psychological virtue requires something else to be realized as politics.

The trouble was that, for my generation of the 1960s, an ordered, focused space like the Pnyx was antidemocratic precisely because it was disciplinary; we believed that freedom lay in breaking the bonds of discipline. Foucault's surgical dissections of disciplinary power frightened us. Moreover, we had an ambivalent relation to linking politics and theater. There was indeed a lot of political street theater in my generation, particularly in protests mocking the Vietnam War. But then, as now, political theater also summoned up the manipulation of public sentiment through clever role-playing, inflamed rhetoric, and artificial scenarios of doom or glory.

These political games might be perennial; they certainly took place in the Athenian Pnyx despite its architectural rigor. Such vices, unfortunately, are abetted by progress in its modern guises. The easy editing of televised imagery, particularly digital images, strengthens the politician's capacity to conceal rather than stand nakedly revealed. Unlike the ancient Pnyx, those watching television's glowing box cannot see each other directly; they rely on what the screen tells them for that sense of polity. It sometimes is said that the Internet might be a new space of democracy, but sociologists tell us that screen communities emphasize denotative statements and short messages. In these communities, the intensity of connection can easily be diminished; to exit from painful confrontation, you need only press a key. Easy, quick decisions are encouraged by such visual conditions, but not the difficult ones requiring time and commitment.

The most urgent social requirement for democratic deliberation today is that people concentrate rather than "surf" social reality. To pay attention and to commit means that our culture needs, in a broad way, to revise its fear of discipline. Indeed, that change occurred in Foucault's own final thoughts about the disciplined care of the self; the polity also requires that care. For this reason, I have come to believe that designers need to pay attention to the architecture of theaters as possible political spaces. Live theater aims at concentrating the attention of those within it. To achieve sustained attention, to commit people to one another even when the going gets rough or becomes boring, and to unpack the meaning of arguments all require a disciplinary space for the eye and the voice.

■ ■ ■

I would like to illustrate the possibility of creating a modern Pnyx by discussing some innovative theater architecture created in the last half century. It is work that addresses, in different ways, how to make an urban theater appropriate for the cities of our time. Even though entirely contemporary in form, these buildings are imbued with the ancient idea that the theater can be used as a space of political congregation.

Perhaps the most innovative is the theater recently created in Tokyo by Japanese architect Tadeo Ando. This is meant as a multiuse space, and Ando's emphasis is how to make speech from the audience as clear as speech from the central stage. Like the ancient Greek theater, Ando's theater uses as much natural light as possible, based on his belief that people can dwell comfortably in a space for longer periods of time in natural light than in artificial illumination.

Although Ando's theater is meant for plays, its other programmed uses include political meetings, and this political program relates perhaps to its most unusual feature. This is a portable theater; it can be taken down and reerected in different parts of the city. Portability has an important political dimension; meetings throughout the city can be organized under common physical conditions, and portability serves a certain equality of discourse.

When we think about the urban dimension of theaters used as meeting spaces, the integration of the theatrical space into the fabric of the city becomes an important consideration. In London's East End, a theater recently has been constructed that attempts this integration both in its siting and in the very articulation of its walls. This is the Angel Theatre. Every window looks and functions like a door. For both plays and community meetings, people walking outside have only to look in to see what is happening, much like the law courts in the ancient agora.

For Americans, these urban theaters might seem alien because so few of us live in the midst of dense cities. The suburban condition is one of dispersion; the densities of the shopping mall or of the big-box store, which keep customers moving rather than sitting and talking, are like crowd islands. In one of his most remarkable late projects, architect Louis Kahn addressed this problem. He sought a theater in which something like a city is contained within the theater's walls. The inner spaces surrounding the auditorium shell are articulated like the streets of an Italian hill town, and the program for this theater imagines these spaces open to the public at all times, even when there are no events in progress. By creating an inner agora, as it were, the program envisions that the theater itself would then become a familiar and natural place in which to hold meetings—large inside the auditorium, smaller in the multiple spaces that traditionally are seen only as foyers.

■ ■ ■

When my team returned from Jerusalem, we tried to make an experiment of our own in political theater. It was an experiment dictated by the site. Outside the Damascus Gate, the triangular area of empty land on which we focused abuts the Arab central business district. Just to the east is the Christian Garden Tomb, meant to commemorate the crucifixion. Next to the Garden Tomb is a Muslim cemetery as well as the remnants of a bus station serving Palestinian East Jerusalem. The triangle itself was, at the time of our journey, filled with buses and parked trucks, overflowing each morning with goods passing through the gate to the old walled city. Modern Jerusalem pressed in on this open triangle—pressed and threatened to explode. This was one of the most hotly contested sites of the new Jerusalem.

Among the plans the Harvard team generated for the Damascus Gate, under the general direction of architect Moshe Safdie, was a conference center fronting a new public plaza. The conference center was in the form of a semi-circular theater meant to be built low so that it would not loom over the walls of the old city. Parking for trucks and a new bus terminal lay tucked beneath an open plaza. This was in many ways a project sensitive to its site; for example, hiding the vegetable and meat trucks below the plaza helped to cope with the intense heat of the sun.

Still, the project lacked the political qualities of the other theaters I have described. A monument to discussion, divorced from the urban fabric of buildings around it, this meeting place did not draw the outside inside. Its open side gave out on an empty space, whereas it should have been turned toward the fabric of streets at its sides or pushed much closer to the masses of people streaming in and out of the Damascus Gate to the old city.

I have come to understand that these limitations, combined with the virtues of the other theaters, suggest one way of answering Nusseibeh's real question: What is urban democracy?

■ ■ ■

In the long course of Western development, democracy has been a relatively rare way of life and a way of life that appeared mostly in cities. Democratic participation has held out the hope of gathering together all people in a city. Ancient Athenians cherished this hope, as did later the citizens of Italian medieval communes and of Reformation

German towns. To realize this hope of coming together, urban democracies sought for a unifying political space to which all citizens could relate—the Pnyx, the parade routes of the communes, the German *Rathaus.* Urban democracy meant centralized power in the sense of a single site, a single image, where all citizens could witness the workings of government.

In the modern era, the hope for democracy has become nearly universal throughout the world, but the nature of democracy that people hope for has changed. National and even global visions of democracy are the old type of urban democracy writ large, a unifying political force. Against those visions has been set another—decentralized democracy, which does not aim at such cohesion. Instead, as the ideal of decentralized democracy first appears in the writings of Tocqueville and Mill, power is portrayed as becoming more democratic in the sense of inviting participation as it becomes more fragmented and partial in form.

Belief in local, decentralized democracy has radical political implications. Taken to the limit, such a belief rejects a single description of the good state or refuses to define citizenship in terms of rights and obligations applicable to each and every citizen in just the same way. Instead, it argues that differences and divergences will develop in practice. The national or global polity will resemble a collage difficult to resolve into a single image.

Decentralized democracy has a particular affinity to the modern city. Cities very rarely are coherent human settlements; that is what Aristotle tried to convey in the term *synoikismos,* a coming together of differences, be they families, economic interests, or political views. In the modern world economy, the fragmentation of urban settlements has increased radically; urban settlements are bigger and more stretched-out places but yet unified.

Decentralized democracy is an attempt to make a political virtue out of this very fragmentation, an attempt that appears in demands for local communal control of schools, welfare services, and building codes. Decentralized democracy also has a visual dimension. This democratic vision may prefer the jumbled, polyglot architecture of neighborhoods to the symbolic statements made by big, central buildings. It may reject the all-at-once, massive development of urban centers such as Berlin's Alexanderplatz and instead seek slower, less coherent growth throughout the city. Ultimately, the result of visual, decentralized democracy should be to shatter those images that attempt to represent the city as a whole.

This is appealing; real life is local, concrete, and particular, but the decentralization of power is in fact not so benign. Gated communities in the American suburbs exercise such local power; communities may decide, by quite democratic means, to exclude blacks, Jews, the elderly, or other "undesirables." Even if the community is benign, the smaller a unit of power, the weaker it becomes. I think, in this regard, of the small communities in upstate New York fighting against IBM during recent years when the giant corporation downsized local workers; the communities are simply too small to fight back effectively.

The word *decentralization* suggests an effort to break up an existing, comprehensive power or to limit its disciplinary tendencies. But as Tocqueville well understood, the process of attacking that central power, breaking it down to ever more local levels, can spin out of control so that ultimately no polity is left at all. In the words of former British Prime Minister Margaret Thatcher, there would remain only "individuals and families," no image of the collective good with which individuals could identify.

This last danger is what theatrical architecture used for political purposes can attempt to combat. Theatrical forms can attempt to develop civic connections not of the fleeting sort, as in a public square, but rather of a more sustained and focused sort. Using theater for this purpose means innovating in its form.

As the designers of London's Angel Theatre understood, a community center for sustained interaction must, in the context of the modern city, be open to casual inspection and entry, as the proposal for Jerusalem's Damascus Gate is not. A good local communal meeting place has to be integrating, especially when a city is fragmenting.

As Ando understood, a portable community meeting place might at least provide common ground in a fragmented city. We never can make do, I believe, with a city whose neighborhoods are identity ghettoes of class or race; the more social isolation, the more possible are violent conflicts or sheer indifference to the fate of others. A portable political architecture, therefore, suggests a way of sharing political activity without unifying it. Ando wants people, as it were, to share a common mental ground in acting locally.

If a public culture is lacking at all in a community, then innovations in theater architecture can at least try to create something out of nothing. This was Kahn's vision of a theater set even in the isolated space of heartland America.

In arguing for the political virtues and design possibilities of theaters, I do not mean that we should forget about building public squares. Because cities gather together differences, strangers need a center; they need somewhere to meet and interact. However, the sheer arousals of the center are not enough to create an urban polity. The polity requires further a place for discipline, focus, and duration; decentralized polities particularly need such places where people can congregate.

Democratic decision making, particularly at the local level, is not fulfilling; local acts cannot realize all that we are capable of imagining about how we ought or want to live. Acting locally in the context of a city entails a loss of coherence, an acceptance of fragmentation. Democracy costs us something psychologically. This is why, in exploring the characteristics of democratic space, I have invoked Nusseibeh's character. Here was a man who saw further than the life of a manager of an ailing electric light company. His wealth and cosmopolitanism would have made it possible for him to have remained in Paris as an exile. Yet, he submitted to the discipline of living locally and so partially. Nusseibeh's sense of the insufficiency of life as we actually manage to live it seems to me relevant in this way to the experience of democracy. In a theater of democracy, his personally unsatisfying relation to others would be shared and sustained; his Israeli captors would share it.

Perhaps this is what I should have replied when he demanded what an urban democracy looks like. He had only to look in a mirror; the answer to his question lay within him.

Index

About the Contributors

ROBERT A. BEAUREGARD is Professor in the Milano Graduate School of Management and Urban Policy at the New School for Social Research in New York City. Currently, he is finishing a historical analysis of the postwar decline of industrial cities in the United States while also writing on planning, urban theory, and urbanist intellectuals. His most recent book is *Voices of Decline: The Postwar Fate of U.S. Cities* (1993), an analysis of the discourse of urban decline.

SOPHIE BODY-GENDROT is Professor of Political Science and American Studies at the Sorbonne-Paris IV and at the Institut d'Etudes Politique in Paris. The former editor-in-chief of the *French Journal of American Studies,* she writes on urban issues, power conflicts, violence, immigrant and minority issues, and social policy from a cross-national perspective. Her most recent books are *Les villes américaines* (1997), *Ville et violence* (1993/1995), *Les Etats-Unis et leurs immigrants* (1992), and *The Social Control of Cities* (in press). She is an author in *The Bubbling Cauldron* (1996) and *Poverty, Inequality, and the Future of Social Policy* (1995).

M. CHRISTINE BOYER is William R. Kenan Jr. Professor at Princeton University in the School of Architecture. She lectures on topics in the history and theory of urbanism and architecture as well as on the effects that computers have on patterns of uneven development. She is the author of *CyberCities* (1995), *The City of Collective Memory* (1993), *Manhattan Manners: Architecture and Style 1850-1890* (1985), and *Dreaming the Rational City: The Myth of American City Planning 1893-1945* (1983).

MICHAEL DEAR is Director of the Southern California Studies Center and Professor of Geography at the University of Southern California. He has been a Guggenheim Fellow and a fellow at Stanford University's Center for Advanced Study in the Behavioral Sciences. He has won awards for creativity in research from the Association of American Geographers and the University of Southern California.

SUSAN S. FAINSTEIN is Professor of Urban Planning and Policy Development at Rutgers University, where she teaches planning theory. Most recently, she coedited (with Scott Campbell) *Readings in Planning Theory* (1996) and *Readings in Urban Theory* (1996) and authored *The City Builders* (1994). She also co-edited (with Dennis Judd) *The Tourist City* (1999).

STEVEN FLUSTY is a doctoral student in the Department of Geography at the University of Southern California, where he uses narrative commodity chains to investigate quotidian globalization. He has worked as a consultant to numerous architects, landscape restorationists, and public agencies. His study of exclusionary urban design was published by the Los Angeles Forum for Architecture and Urban Design under the verbose title "Building Paranoia: The Proliferation of Interdictory Space and the Erosion of Spatial Justice."

ALAN MABIN has traveled and researched widely in Australia, Brazil, Canada, Kenya, Malaysia, Namibia, the United States, and Zimbabwe. He has held fellowships at Yale University and the Université de Paris-X and is the editor of three books and author of many articles in the fields of urban planning, economic history, and urban geography. He currently is Associate Professor in Town and Regional Planning and Director of the Centre for Development of the Built Environment at Witwatersrand University (Johannesburg). He also is Deputy Chairperson of the South African Development and Planning Commission and chairs the board of directors of a housing nongovernmental organization in Johannesburg.

GUIDO MARTINOTTI, Harkness Fellow (1962-1964) with postgraduate studies in sociology at Columbia University and the University of California, Berkeley, is Chair of *Sociologia urbana* at the Università degli studi di Milano and is a member of the Committee for the Creation of the Faculty of Sociology in the new Second State University of

Milano. From 1992 to 1996, he was chairman of the Standing Committee for the Social Sciences of the European Science Foundation and recently was reelected as a member of the European Science and Technology Assembly. In 1998, he was a Fellow at the Remarque Institute at New York University. Also in 1998, he was charged by the Italian minister of university research and technology to coordinate the committee entrusted with the reform of the Italian University. Since 1989, he has been a member of the Academia Europaea and Convenor for Sociology. His most recent publication is *Perceiving, Conceiving, Achieving the Sustainable City: A Synthesis Report* (1997) for the European Foundation for the Improvement of Living and Working Conditions. He also has published *Cittadini si diventa* (1996), *Metropoli* (1993), and *Informazione e sapere* (1992) and has coedited *Bisogni informativi, banche dati e territoria* (1994) and *Education in a Changing Society* (1978).

MARGIT MAYER teaches comparative and North American politics at the Free University of Berlin. She received her Ph.D. and Habilitation degrees from the University of Frankfurt/Main. Among her research fields are urban politics and social movements. She also engages actively in the local movement scene by collaborating with groups in Berlin and, when the opportunity arises, in North American cities. She has co-authored *Das neue Gesicht der Städte: Theoretische Ansätze und empirische Befunde* (1990) (with S. Krätke, R. Borst, R. Roth, and F. Schmoll), *Politik in europäischen Städten: Fallstudien zur Bedeutung lokaler Politik* (1993) (with H. Heinelt), and *The German Greens: Paradox Between Movement and Party* (1998) (with J. Ely). She currently is working on a book on urban social movements.

THIERRY PAQUOT teaches philosophy at the Ecole d'Architecture of Paris-La Défense (Nanterre) and comparative history of urban civilizations at the Ecole Nationale des Ponts et Chaussées. He supervised the publication of the collection of essays *Le monde des villes* (1996) and has published numerous essays on the urban question including *Homo urbanus* (1990), *Villes et civilization urbaine XVIIIe-XXe siècles* (1992), *Viva la ville!* (1994), and *L'Utopie ou l'idéal piège* (1996). He edits the review *Urbanisme, le magazine international de l'architecture et de la ville.*

CHRISTIAN RUBY is a philosopher who lives in Paris. His recent publications include *L'Enthousiasme: Essai sur le sentiment en politique* (1997) and *La Solidarité: Essai sur une autre culture politique dans un monde postmoderne* (1997).

SASKIA SASSEN is Professor of Sociology at the University of Chicago. Her most recent books are *Globalization and Its Discontents: Selected Essays 1984-1998* (1998) and *Losing Control? Sovereignty in an Age of Globalization* (1996). Her books have been translated into several languages. *The Global City* recently has appeared in French, Italian, and Spanish. She currently is completing *Immigration Policy in the Global Economy: From National Crisis to Multilateral Management* and has begun a new project on "Cities and Their Cross-Border Networks."

RICHARD SENNETT is University Professor of the Humanities at New York University and Centennial Professor of Sociology at the London School of Economics. He has published numerous books on cities and urbanism including *The Uses of Disorder* (1970), *The Fall of Public Man* (1974), and *Flesh and Stone: The Body and the City in Western Civilization* (1994). His newest work is *The Corrosion of Character: The Personal Consequences of Work in the New Capitalism* (1998).

MICHAEL PETER SMITH is Professor of Community Studies and Development at the University of California, Davis, and Chair of the Graduate Group in Community Development. Prior to his appointment at UC Davis in 1986, he was Professor of Political Science at Tulane University, where he taught urban politics for 12 years. He also has taught and researched at Dartmouth College and the universities of Cambridge, Essex, California (Berkeley), North Carolina, and Boston. In the fall of 1998, he was a Visiting Scholar at the International Center for Advanced Studies at New York University. An internationally recognized urban theorist, he has published 18 books, including *The City and Social Theory* (1979/1980), *Cities in Transformation* (1984), *The Capitalist City* (1987), *City, State, and Market* (1988), *The Bubbling Cauldron* (1995), and *Transnationalism From Below* (1998). He currently is writing a new book on transnational urbanism.

NEIL SMITH teaches geography at Rutgers University and is a Senior Fellow at the Center for the Critical Analysis of Contemporary Culture. His books include *The New Urban Frontier: Gentrification and the Revanchist City* (1996) and *Uneven Development: Nature, Capital, and the Production of Space* (1984). He currently is completing a study of "The Geographical Pivot of History: Isaiah Bowman and the Geography of the American Century."